UNDERPINNING

UNDERPINNING

A PRACTICAL GUIDE

Roger A. Bullivant
and
H.W. Bradbury

Blackwell
Science

© 1996 by
Blackwell Science Ltd
Editorial Offices:
Osney Mead, Oxford OX2 0EL
25 John Street, London WC1N 2BL
23 Ainslie Place, Edinburgh EH3 6AJ
238 Main Street, Cambridge
  Massachusetts 02142, USA
54 University Street, Carlton
  Victoria 3053, Australia

Other Editorial Offices:
Arnette Blackwell SA
224 Boulevard Saint Germain
75007 Paris
France

Blackwell Wissenschafts-Verlag GmbH
Kurfürstendamm 57
10707 Berlin, Germany

Zehetnergasse 6
A-1140 Wien, Austria

All rights reserved. No part of this
publication may be reproduced, stored in a
retrieval system, or transmitted, in any
form or by any means, electronic,
mechanical, photocopying, recording or
otherwise, except as permitted by the UK
Copyright, Designs and Patents Act 1988,
without the prior permission of the
copyright owner.

First published 1996

Set in 10/13 Times
by DP Photosetting, Aylesbury, Bucks
Printed and bound in Great Britain
by Hartnolls Ltd, Bodmin, Cornwall

DISTRIBUTORS

Marston Book Services Ltd
PO Box 87
Oxford OX2 0DT
(*Orders*: Tel: 01865 791155
         Fax: 01865 791927
         Telex: 837515)

USA
Blackwell Science, Inc.
238 Main Street
Cambridge, MA 02142
(*Orders*: Tel: 800 215-1000
         617 876-7000
         Fax: 617 492-5263)

Canada
Copp Clark, Ltd
2775 Matheson Blvd East
Mississauga, Ontario
Canada, LW4 4P7
(*Orders*: Tel: 800 263-4374
         905 238-6074)

Australia
Blackwell Science Pty Ltd
54 University Street
Carlton, Victoria 3053
(*Orders*: Tel: 03 9347-0300
         Fax: 03 9349-3016)

A catalogue record for this title
is available from the British Library

ISBN 0-632-04004-1

Library of Congress
Cataloging-in-Publication Data
is available

Contents

Preface	ix
Glossary of terms	xi

Chapter 1	**The development of underpinning**	**1**
	Introduction	1
	Movement effects	4
	Structural behaviour of buildings	4
	Buildings likely to need underpinning	6
	Categories of underpinning	7
	Types of foundations	8
	Growth in underpinning	12
Chapter 2	**Causes and indications of damage**	**14**
	Settlement and heave	14
	Mining subsidence	20
	Erosion, slope movement, ground decomposition and vibration	23
	Trees and vegetation	24
	Cracks and cracking as damage indicators	27
	Assessing damage by crack measurement	30
	Repairs to cracks	34
Chapter 3	**The survey process and site investigation**	**37**
	Initial assessment of damage	37
	Methods of investigation	38
	Site and soil investigation	38
	The survey process	39
	The site assessment	40
	Ground investigation	42

Whether or not to underpin	46
What will a site investigation achieve?	48
Conclusions	49

Chapter 4 Traditional underpinning — **50**
Mass concrete systems	50
Pad and beam (beam and pier) systems	54

Chapter 5 Piling in underpinning — **59**
Piling techniques	60
Mini piling	70
Specifications for underpinning work	75
Infilling of piles	77
Summary of piling solutions and usage	79

Chapter 6 Pile design and piling equipment — **83**
Behavioural determinants of pile design	83
Example results and calculations	86
Plant and equipment	87

Chapter 7 Grouts and grouting — **94**
Ground improvement by grouting	94
Selecting a grout system	95
Cementitious grout blends	97
Chemical grouts	98
Flowing concrete	98

Chapter 8 Safety and legal aspects of underpinning — **100**
Safety	100
The CDM Regulations	101
Difficulties associated with underpinning	103
Main legislation affecting underpinning work	104
Temporary supports for premises being underpinned	105
Approval of underpinning designs	109
Guarantees	110
Contracts	112
Civil liability for foundation failure	113
Control of Pollution Act 1974	115

Chapter 9 Customer care and quality assurance — **116**
Householders' concerns	117
Insurer's viewpoint	118

Handling underpinning projects	118
Quality	121
Accreditation	123

Chapter 10 Prevention is better than cure — **125**

Good practice for prevention	125
Hazards and precautions	128
Ground improvement techniques	131
Innovative foundation systems	136

Appendix A Case studies	*143*
Appendix B Comparative costs	*169*
Further reading	*171*
Index	*175*

Preface

This book is intended as an overview of underpinning for domestic and light industrial buildings and covers work normally undertaken by building and civil engineering contractors, and specialist underpinners. It provides a one-stop reference point covering most aspects of underpinning work and will prove useful to structural and design engineers. Underpinning requires expertise in the execution, along with safe working practices. These are discussed in the book and guidance offered.

When considering settlement and subsidence of foundations it should be borne in mind that the real foundation of any structure is the ground beneath it, and the concrete, raft or piled foundations are merely means of transferring the load of the structure on to acceptable soil strata. Thus it is the ground which fails and not the man-made support, given good workmanship and materials.

In domestic buildings where ground failure has occurred resulting in structural damage, underpinning to restore stability followed by structural repairs often provides the optimum solution. The selection of the type of underpinning available and the steps leading to such choice are discussed. In addition to remedial work, underpinning techniques also offer a solution to building new foundations in difficult soil conditions.

The book sets out many aspects of selection and construction, as well as the steps involved in underpinning. It is hoped it will make a positive contribution to good practice and will also widen the knowledge of all concerned, from client and architect to loss adjusters, insurance companies, building societies, contractors and engineers. It may also be of help to householders faced with the problem of structural failure.

R.A. Bullivant
H.W. Bradbury

Glossary of terms

Allowable bearing pressure – the maximum allowable net internal load of the soil at the underside of the foundation, after allowing for the ultimate bearing capacity, the degree and type of settlement anticipated and the factor of safety required.

Compaction – the increase in strength and density of a soil generated by impact.

Compressive deformation – settlement without upthrust as the soil particles are compressed together causing compaction.

Consolidation – the increase in the density and strength of soil caused by the expulsion of water an/or increased loading.

Density – weight per unit volume. Bulk density is the total weight, including contained water, divided by volume. Dry density is the weight of solids divided by total volume.

Elastic deformation – a temporary lateral upthrust of ground which returns to normal on removal of loading.

Heave – an upward or lateral movement of ground by swelling of the strata due to a variety of causes. Heave is also associated with shrinkage, usually as a result of variations in moisture content, and can be cyclical in nature due to seasonal changes, ground recovery, or changes in adjacent vegetation.

Land slip – a sudden movement on a slope whereby the soil creeps or moves downwards over time.

Liquid limit – the minimum moisture content at which soil will flow under its own weight.

Piles – usually long slender concrete or steel members for transferring foundation loads through incompetent strata to safer deeper high bearing compacted ground.

Plastic deformation – one which causes a permanent deformation accompanied by lateral upthrust. It is caused by lateral flow of a soil under load.

Plastic limit – the minimum moisture content at which any further reduction in water content will not cause a decrease in volume.

Settlement – a downward movement of the ground and/or a structure thereon, due to loading. This movement may cause cracking and distortion.

Subsidence – a downward movement of the ground caused by the effects of mining, washout or subterranean excavation (i.e. a volumetric change in the substrata). It can be universal or differential and is likely to occur on a site even without applied loadings.

Underpinning – is a branch of foundation engineering concerned with remedying inadequate ground support to a structure, due to a variety of causes, but without the necessity of removing the original foundation or damaging further the superstructure. The object of underpinning is to safely transfer the foundation load from its original bearing level down to a new, lower, and safer level composed of more stable, competent subsoils.

Ultimate bearing capacity – the net value at which the ground fails internally in shear.

Chapter 1
The development of underpinning

Introduction

Casual observers could be forgiven for assuming that the alteration in the level of buildings caused by ground movement is a relatively recent phenomenon. Prior to 1971, the majority of repair work to be seen would have been householders tackling as routine maintenance the telltale signs of such damage in domestic buildings – cracks in walls and ceilings, and doors and windows that stick. Their efforts would have been restricted to how much they could afford and how well they understood the symptoms. However, since 1971, the effects of ground movements are covered by insurance and it is now common to see specialists carrying out substantial underpinning work to rectify, rather than disguise, the problem.

The problem of settlement has been with us since the earliest times. Prehistoric people probably attempted to prolong the lives of their dwellings by ramming earth and stones under the foundations. Timber was used in many early domestic buildings and later it provided the material for a nascent form of underpinning (as seen in Roman fascines, for example). During these times, little science was involved in the design of foundations. The knowledge that existed about soils and structures had been gleaned from local experience and the skills handed down by artisans.

The application of a scientific approach to foundation work began in the seventeenth century and once lime came into use as the matrix for mortars and concretes it became practicable to improve defective foundations. However, when Joseph Aspdin of Leeds patented Portland cement in 1824 it became apparent that materials technology was progressing faster than soil mechanics and foundation theory. Engineers then recognised that the stresses induced in subsoils by excavation and loading were significant and that, since strain could be related to deformation, the deformed shape of the structure reflected elastic and plastic distress. The theories of elasticity and plasticity were subsequently evolved and, by the mid-1920s, the mechanics of soil behaviour and their relationship to fluid pore pressures and volumetric change became understood. During the same period, engineers realised that theory and practice both have roles to play in good design and, along with field observations, lead to validation of theoretical principles. For instance, the characteristics of soils in situ differ from those of the same soils under laboratory conditions so correlation is necessary in order to draw valid conclusions.

This marriage of scientific theory and practical knowledge was initially exploited in the construction of large multistorey structures. Low rise buildings did not warrant the same sophisticated technical and theoretical analysis. Of primary importance for these projects were the techniques and expertise derived directly from site experience and practical knowledge of the ground. Although modern low rise developments now benefit from the latest design techniques there is still a big difference between constructing a new house and strengthening one already built.

In an existing building suffering ground movement difficulties, the settlement or heave affecting the foundations and any ensuing damage to the upper walls are aspects of the same problem and should be viewed together. This may require advice from both the structural and geological experts. A new building, on the other hand, can be designed from the outset to satisfy likely site conditions. The age of the building is also relevant. An old building which has stood securely on its foundations for a long period can provide useful data to the experienced observer and thus warrants a careful examination. However, this can give rise to differing opinions on what method of restitution work should be adopted. Whichever system is used, the introduction of further stresses and strains into the structure can lead to a serious alteration of a building's existing equilibrium. It is therefore necessary to consider very carefully the causes of movement and structural behaviour of the dwelling before embarking on any repair solution.

Generally in the past, repair work to foundations was entrusted to craftsmen and was limited to the means available to them. Almost exclusively, the work involved techniques that can be grouped under the heading 'traditional underpinning'. Traditional underpinning using mass concrete is still used today, but it is expensive, labour intensive, and disruptive. It transfers the load of the existing foundation to a lower and hopefully more suitable load bearing strata, so stabilising shallow foundations. It involves excavation to a depth normally not exceeding 1.5 m, and infilling the void created with concrete. This method accounts for a great deal of underpinning work at present and is often carried out by local builders. The traditional approach has limitations and relies often on the arbitrary approval of an excavated base, usually without the benefit of an adequate soil investigation. It is thus a risky process and is being superseded gradually by the pad and beam system, as described later, and by more sophisticated underpinning methods.

With the climatic change now taking place (overall warming due to the greenhouse effect) and the burgeoning effects of modern-day tree growth or removals, it is often considered that 'trad' underpinning is unsuitable and no longer cost effective, particularly in clays. The depth of potential soil desiccation is often now found at 3–5 m in Europe and much deeper in parts of the USA and the Middle East. This depth can be even greater in areas affected by water abstraction, sewerage/drainage works, deep excavations, and tree root infestation as trees have to seek water at ever increasing depths. Only in the last few decades has underpinning work shifted away from general building contractors to specialist companies. The growth in this field has also meant that specialist underpinning work is now a realistic economic option for home owners.

More technically orientated systems of underpinning have taken over from mass concrete traditional systems as a result of greater understanding of the theory of structures and the behaviour of soils. Structural and value engineering now produce designs that can cope safely with the forces to which buildings are subject, and, in the case of underpinning, effective designs are available to restore stability to affected dwellings. A building can be considered stable if a change in its form or condition will not cause even partial collapse and the structure will withstand any displacements caused by normal loadings in use, along with changes resulting from anything other than major accidents.

All structures are subject to some movement during construction and certainly afterwards when in service. Furthermore alterations to a building can reduce its stability as may changes in its use, or alteration in prevailing ground conditions. Movements occur where loadings, temperature and/or ground conditions alter. Initially it may appear logical to strengthen the foundations before repairing the building itself. However, this is not always the case. It is often advisable to strengthen the structure above the ground, either by permanent repairs or strutting and shoring (temporary work), before carrying out any underpinning. Experience or prudence will indicate the necessary precautions.

When building a new structure the functions of the architect and the engineer are usually quite distinct and separate. However, the task of underpinning an existing foundation requires a close relationship between all parties and it is therefore advisable to employ specialist firms for works of restoration and underpinning. Such companies have all the relevant skills or are accustomed to working in partnerships and so are in a position to both design and execute the works competently. Having said that, between 50% and 70% of underpinning work is still done by building contractors using the traditional mass method.

Movement effects

There are three main causes of structural movement in buildings:

(1) *Loadings* Loading through day-to-day use can result in bending and shear deformation, strain, settlement, consolidation or subsidence.
(2) *Temperature* Fluctuation in temperature can result in expansion, contraction and bending movements.
(3) *Ground moisture* Variations in moisture content can cause shrinkage, expansion, heave or swelling and erosion, particularly in clay soils.

When movements from whatever cause are restrained or differential in effect, deterioration of the structure can result in a loss of stability. Where the movements are progressive, the accumulation of defects may become irreversible. When rotational movement occurs, it can cause damage in floors and roofs as well as interior and external walls indicated by cracking. Any or all of the foregoing in a building *may* indicate a necessity for underpinning work.

Structural behaviour of buildings

A building is invariably affected by slight movement and damage during construction and for some time after occupation as the premises settle into the underlying strata. This initial settlement is usually accompanied by fine cracking. This can be repaired easily and is classed as 'cosmetic damage' because it does not affect the structural integrity of the building.

The structural behaviour of a dwelling must be examined in its entirety to understand how foundations behave when a downward movement of the foundation occurs. Excluding purely structural failures or roof spread, a downward vertical movement of a wall involves a rotation of the foundation about its base resulting in a partial loss of stability within the structure. If the movement is excessive it may even cause collapse. Once identified, preferably at an early stage, this can be counteracted by underpinning, that is, the provision of a support under the foundations and ideally within the footprint of the wall (invariably that is within the middle third of the foundation as is shown in zone A of Fig. 1.1). Thus when the foundations of a domestic dwelling fail due to settlement or consolidation of the ground below, observation has confirmed a tendency for the outer walls to rotate (i.e. to move downwards and outwards to varying degrees). The walls are normally restrained from inward movement by the floors and partitions.

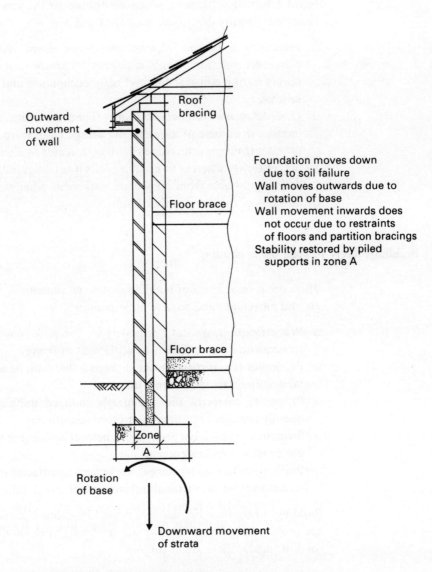

Fig. 1.1 Structural behaviour of a dwelling.

Movement is greatest at the outer edge of the wall and additional support is therefore indicated. Such movement may be immediate or progressive. Once the mode of failure has been determined an experienced foundations engineer can design a solution for a return to stability and equilibrium. This usually involves some form of underpinning.

Successful underpinning requires an understanding of many disciplines such as the materials sciences, geology, structural engineering and, most important, experience and flair. Buildings are subject to different kinds of settlement which contribute to the general structural behaviour already described. Broadly, these are:

- *Immediate settlement* A combination of elastic components and plastic deformation without change in volume or water content. This occurs during construction and early occupation and is usually not serious.
- *Consolidation settlement and heave* These are due to a reduction or increase in volume of the substrata caused by the removal or ingress of water from the soil voids as loading increases or alters. In clays this occurs slowly whereas in granular soils it occurs rapidly and is often indistinguishable from immediate settlement other than by the seriousness of the damage that results.

Buildings likely to need underpinning

There are several kinds of building that could potentially be affected by ground movement and require underpinning:

- Properties that are not adequately or correctly founded and have subsequently been affected by settlement or heave.
- Properties founded on unstable ground that will, as a result, inevitably suffer damage at some time.
- Properties correctly and adequately founded initially, but subsequently damaged by climatic or other conditions.
- Structures which have been or will be overloaded due to a change of use or partial reconstruction.
- Buildings affected by external works such as adjacent excavations or vibration from heavy manufacturing plant.

Buildings likely to suffer damage caused by ground movement also fit age profiles that dictate the way underpinning work should be approached. Roughly, these are as follows:

- *Very old or 'ancient' buildings, usually 'listed' (i.e. recognised as having special architectural or historical interest and therefore protected from demolition or alteration), and generally older than 150 years of age.*

Considerable care is required to stiffen, grout, shore strut or pre-repair these old buildings before attempting any underpinning work. Strict controls on works on such buildings are in existence and consultation, advice and permissions required.

❏ *Rather more 'recent' buildings – completed between 50 and 150 years ago.*

Great care is still required, but generally the structures have a greater amount of structural integrity. Some have the protection of being listed.

❏ *Current/modern buildings – under 50 years old.*

Generally these buildings have been constructed in accordance with some form of Building Control Regulations or Codes of Practice. Usually they are reasonably well constructed and have a good degree of structural stability.

Categories of underpinning

There are basically four general categories of underpinning. These are:

(1) *Remedial* Underpinning for remedial purposes to restore stability following ground movement.
(2) *Conversion* Underpinning with the intention of permitting increased loading on the foundations such as a vertical extension or structural alteration.
(3) *Protective* Underpinning to maintain rights of support, or to protect a building during construction works for a basement, an underground car park, or a swimming pool, or to protect it against external adjacent works such as tunnelling, excavations, vibrations etc.
(4) *Mining* Underpinning to remedy the deleterious effects of mining subsidence. This may involve the provision of a jacking system to provide adjustment for ongoing future movements after mining has ceased, or as advised by a mining engineer. British Coal (previously the National Coal Board) have collated all coal and oil shale mining records and these are available for inspection or copying on payment of a fee. For abandoned mines and quarries covering other minerals Her Majesty's Inspectorate of Mines has now passed all records back to local authorities. These are held by a variety of departments but primarily by county archivists. The Land Valua-

tion Department of the Inland Revenue can also be a useful source of information. The system is now similar to that in the USA where private companies and state authorities hold the relevant records.

All categories require the use of similar systems but for different reasons and to different design criteria. Within these groupings underpinning has grown into a significant international industry during the past 20 years.

Types of foundations

It is also important to bear in mind the types of foundation likely to be encountered in domestic low rise buildings in order to be able to arrive at an appropriate underpinning remedial solution when the need arises.

The purpose of a foundation is to provide a firm, stable base on which to build a structure in the light of prevailing ground conditions by transferring the weight of the structure onto the underlying ground in such a way as not to cause excessive soil distress and movement.

The foundation types most likely to be encountered in the UK in low rise buildings are:

- Continuous strip foundations or
- Monolithic slab or raft foundations.

In the USA, because of differing construction techniques, there are in addition:

- Isolated spread footings;
- Drilled shaft, pier, bored pile or cast in place uncased pile foundations; and
- Down pile foundations.

The latter two are usually used in conjunction with grade beam systems.

Continuous strip foundations (Fig. 1.2)

In the UK, the concrete foundations may be reinforced or unreinforced, the latter being most common. In older buildings the brick footings may be spread in 57 mm steps, as shown in Fig. 1.2. In buildings constructed during the past 50 years, the brick footings have been dispensed with and the wall sits directly on the concrete, which is deeper than in most older buildings and subject to a minimum depth laid down in the Building Regulations (usually 150 mm). However, instead of the recommended minimum of 150 mm, good practice often involves using depths of

The development of underpinning 9

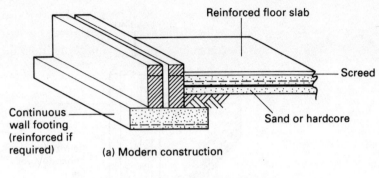

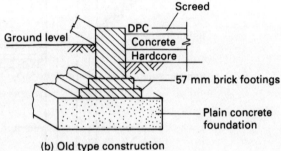

Fig. 1.2 Continuous strip foundations

230 mm for low rise domestic buildings to allow for differing subgrades. The width of the footings today usually extends 150 mm from each edge of the brickwork, giving a width of concrete foundations equal to the width of the wall plus 300 mm. However, where the bearing capacity of the ground and the superimposed load are known the widths can be calculated mathematically as referred to elsewhere in this chapter.

Monolithic slab or raft foundation (Fig. 1.3)

This is simply a reinforced concrete slab sitting on a sand cushion with thickened edges, and may be strengthened by internal integral beams or rafters. Design recommendations can be found in British Standards and the Building Regulations.

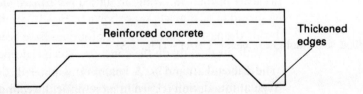

Fig. 1.3 Monolithic slab or raft foundation

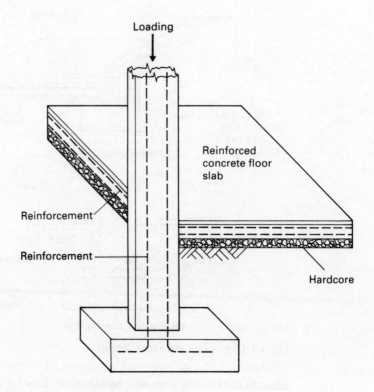

Fig. 1.4 Isolated spread footings

Isolated spread footings (Fig. 1.4)

These consist of square, rectangular or circular concrete bases beneath stanchions or load bearing columns. They are constructed by excavating or drilling into the soil to designed dimensions, inserting steel reinforcement and then pouring concrete. The footing increases the contact area between soil and column, thus reducing any transferred stress.

Example A 12 inch square column designed to carry a load of 10 000 lbs and placed directly onto the soil would exert a contact pressure of 10 000 lbs per square foot. If placed on a 24 inch square base of sufficient depth to eliminate punch shear, the contact pressure is reduced by a factor of four and is thus 2500 lbs per square foot.

Drilled shaft and grade beam (USA) – (Fig. 1.5)

The drilled shaft and grade beam is encountered mainly in the USA. This type of foundation is used in more difficult ground. Support is obtained from skin friction between the shaft and the ground plus an end bearing

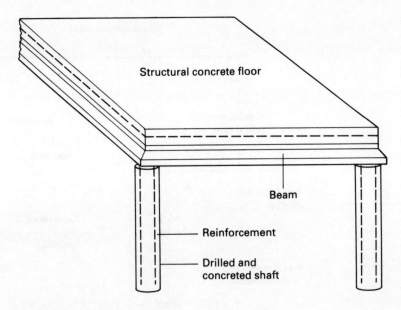

Fig. 1.5 Drilled shaft beam

contribution. To improve the carrying capacity, bell shaped bottoms are sometimes found. In the case of expansive clays, shafts necessarily pass through them and are socketed into non-expansive strata.

Driven piles

These are precast concrete, timber or steel tube piles, relatively long, and driven deeply into the soil to provide the necessary support. They are referred to later when used in underpinning. In the USA the piles can be found singly or in groups, and have pile caps cast on their tops to receive ground beams etc.

Driven piles are not usual in the initial construction of low rise building but passing reference is included here for the sake of completeness. However, in the future, piled foundations in conjunction with precast concrete beam foundations may well come into general use in suspect ground. They are also used increasingly in underpinning work. It is useful to have some idea of the types of pile available to the underpinner. Figure 1.6 shows those types which are normally included under the heading of displacement piles.

12 Underpinning

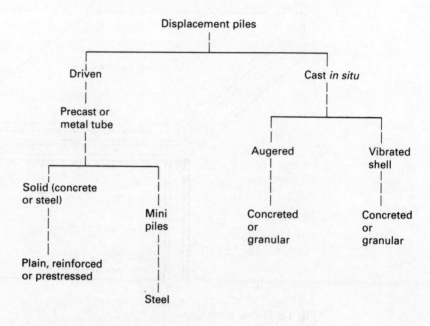

Fig. 1.6 Types of displacement piles

Growth in underpinning

The demand for underpinning has increased steadily as renewals and refurbishment work have gained popularity and as more and more buildings have been constructed on substandard land because the best sites have already been developed or have become prohibitively expensive.

Insurance claims for settlement and/or heave damage to domestic properties have increased significantly in the UK and appear to be following a similar trend world-wide. Insurance for domestic properties generally now includes cover for subsidence damage and this, along with the increase in home ownership, has boosted the demand for remedial repairs arising from suspected failure of foundations. In turn, the demand for underpinning has also increased.

Insurance and the growth in domestic underpinning

In the 1960s the protection provided in the normal household insurance policy against 'subsidence' damage was severely restricted. In the UK in 1971 policies started to give unqualified cover against such damage. By

1975 in the UK the number of claims had risen steadily to about £5 million per annum. In the intervening years up to 1981 claims diminished and then rose rapidly to a peak in 1984 of in excess of £200 million from some 24 000 claims. This was followed by a further slight reduction up to 1988, followed by a very rapid increase during the year 1989 to 26 000 claims costing about £230 million. This increase persisted during 1990 and by 1991 had reached £545 million and 52 000 claims. Because of the recession and a hardening of attitudes by loss adjusters and insurers the level of claims has now been reduced. The reasons for the wide fluctuations is due to the very dry weather experienced, particularly from 1988 onwards. Often today a householder is asked to provide an agreed sum on every claim before the insurance company takes over – a figure of £5000 or more is not uncommon in the UK. Clients in their policies should ascertain that the insurers accept an underpinning solution as curing the problem of movement and that they will maintain the original or acceptable level of insurance cover for the future without undue increases in premiums.

In the UK there has never been any benchmark against which property owners, building societies, insurance companies, loss adjusters and consulting engineers could assess the professional and technical competence of underpinning contractors. There now exists an Association of Specialist Underpinning Contractors formed to provide such accreditation and to provide some assurances as to the resources in finance, plant and people to carry out underpinning work in a satisfactory and competent manner. It also aims to liaise with other professional organisations to enhance the knowledge base and science of underpinning.

Chapter 2
Causes and indications of damage

The areas most likely to be affected by foundation movement are areas founded on shallow clays, areas where trees have been felled, areas where new trees have been planted, areas affected by deep mining, areas with unstable subsoils (i.e. those containing peat, water bearing silt or running sand), and filled sites such as old quarry workings.

In the UK the clay areas are those founded on the Gault, Reading, Oxford, Kimmeridge, Lias, Weald, Boulder and London clays. The desiccation of clay soils will cause ground movement due to ground shrinkage. Eighty per cent of all underpinning work tends to be located south of Birmingham, corresponding to areas with predominantly clay soils. Figure 2.1 shows the distribution of work.

Settlement and heave

All structures settle under load, and naturally 'take up' their new position over many months or years after construction. Further settlement of foundations is caused by stresses generated in the subsoils, or movement related to desiccation. In clay subsoils, the settlement tends to occur over long periods of time and is referred to as progressive. It may show itself as settlement (or heave – see below), depending on the moisture content (see Figs. 2.2 and 2.3).

In the case of granular subsoils, settlement under loading tends to be very rapid. Settlement only becomes problematical when it becomes excessive, or irregular.

Damage ensues as a result of deformation of the building arising from differential movements caused by variations in loading of the structure, or in the variable bearing capacity and characteristics of the soils beneath the structure. The adverse effect of traffic or adjacent building work can also induce differential settlement.

In cohesive soils, changes in moisture content can cause swelling or

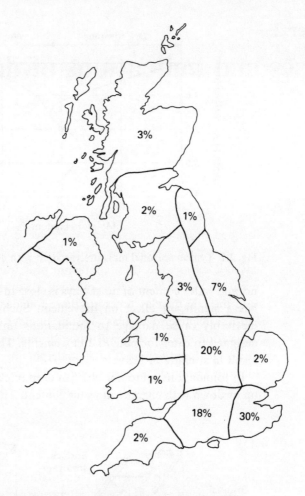

Fig. 2.1 Distribution of underpinning work in the UK

shrinking, both of which may be multidirectional. Heave therefore is expansion or swelling and the opposite of settlement or shrinkage. Both heave and settlement occur in clay soils. A clay soil is one which is composed entirely of fine clay mineral particles below 0.002 mm in size.

Heave is a complex problem associated with prior shrinkage. This prior shrinkage can arise from interruptions of available water supply caused by such things as tree and vegetative growth, water abstraction by pumping or other diversions of ground waters. In clay soils which have low permeability due to their flat plate shape, and affinity for adsorbed moisture, volume can vary considerably with moisture change. Clay soil shrinks as it dries out (desiccates), causing settlement, and swells when moisture returns (rehydration or wetting), causing heave. Since the

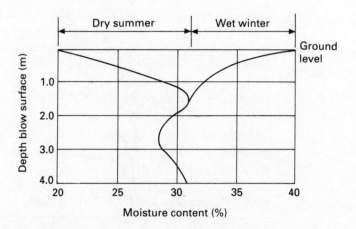

Fig. 2.2 Typical seasonal moisture variation with depth

normal water content of most clays is close to a stable level, desiccation has a significant effect on movement. Such changes in volume will inevitably cause damage to foundations unless they are specifically designed to cater for potential movements. The return of moisture will result in swelling or heave.

In foundations up to about 1.5 m deep in clay, the ground will move up or down with changes in water content. The actual movement range

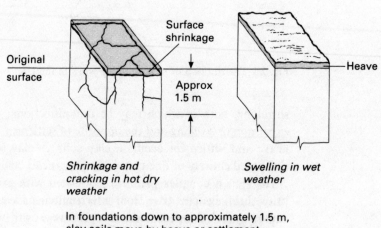

Fig. 2.3 Clay movement

should be assessed over time (spanning at least the extremes of winter and summer) if the condition of the building so permits. (See Figs. 2.2 and 2.3). Alternatively an indirect assessment can be made by the clay content of the soil expressed as a percentage of the whole soil, and the plasticity indices. However, the results can vary with the types of clay, and the clay content of the soil, and do not fit easily into a firm classification system. Thus a degree of judgement is necessary in deciding on the likely extent of movement. Table 2.1 gives a guide to the movement potential of clay soils. Although it is not precise, it provides a useful approximation.

Table 2.1 Guide to movement potential of clay soils

Clay fraction (% of total soil)	Plasticity index (liquid limit – plastic limit)	Movement potential (settlement or heave)
60	35–60	Very high potential
70	60–70	
Generally greater than 60	Generally greater than 35	
55	30	High
60	40	
70	45	
80–90	50	
Generally between 55 and 90	Generally 25–50	
15	12	Medium
20	23	
30	32	
Generally below 30	Generally 12–32	
Below 20	Below 12	Low

Where clay heave is anticipated the use of a designed pile system for remedial works (underpinning) can reduce further damage. Piles should have a sufficient proportion of their length installed into the clay to resist uplift where this exceeds the compression in the piles by reason of the wall loads. If solid rock is founded below the clay, the pile should be located into a rock pocket and movement will then not occur.

Heave movement can be reduced by using a slip membrane such as a pvc tube for short bored piles or a resin or bituminous coating on precast piles at least on the top 1–2 m depth of pile. Alternatively where a precast concrete ground beam is used, a void can be formed under the suspended floor slab and a compressible filler board installed on the outside of the beam. It should be noted as an indication of heave that in clay soils observation of the surface will often show surface cracks up to 25 mm wide by up to half a metre deep or thereabouts. These will be present only during dry periods and in the summer. (See Figs. 2.4 and 2.5.)

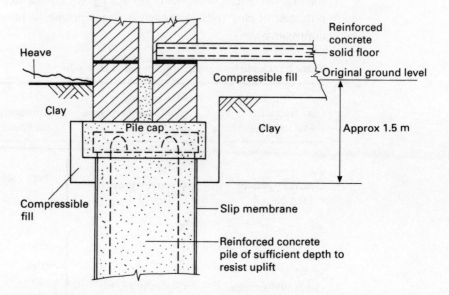

Fig. 2.4 A suggested 'heave' design in clay

Desiccation is a common problem with soils which shrink when they are dried, and which when wetted swell and generate ground heave. The desiccation zone within clays can extend in extreme cases down to 6 m or more, but usually it extends to 2–3 m deep. Desiccation can be identified easily during a site investigation. First, when the moisture content is low, the measured strength is high. The mention of live roots in the borehole descriptions is also an indication of the extent of desiccation.

When the effects of desiccation reach 3 m, the most commonly adopted foundation solution is to use piles penetrating the desiccated soils and founding within the soils beneath. The vertical piles will also be subjected to swelling uplift. Hence, they must be designed with sufficient depth of penetration below the desiccated zone to develop enough skin friction to resist the uplift forces.

In domestic properties heave is generally more serious than shrinkage settlement. This is because in shrinkage clay tends to move away from

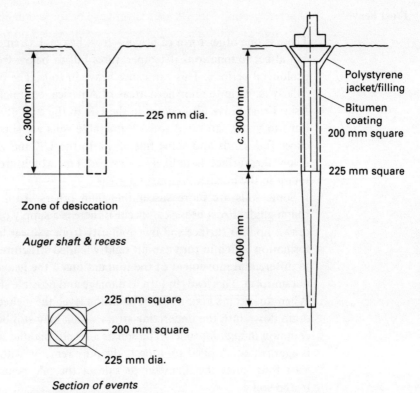

Fig. 2.5 Heave solution using precast piles

the foundations and the attendant strain in the soils is therefore not imposed directly onto the foundation. In a heave situation the clay expands and moves towards the foundation, applying pressure in all directions. Because downward movement is generally more restrained, the worst effects arise from vertical and horizontal displacements. The forces are such that they cannot be wholly contained and will invariably damage the foundations. Special design criteria need to be met so as to permit free movement of the ground without affecting the foundations.

Frost heave

There is one other form of heave – frost heave. Extremely cold weather can affect foundations if temperatures fall to below freezing point for prolonged periods. This can cause heave in soils. The problem is particularly acute in the northern areas of America, continental Europe and Asia. Frost heave also occurs in the UK in the case of shallow foundations in water saturated soils. Vulnerable soils are silts, chalks, coarse clays, fine sands and some fine clays. In the UK the vulnerable depth below the surface is unlikely to exceed 1 m. Much greater depths are found in the norther continental areas.

Some soils are more susceptible than others. Table 2.2 presents a rough guide. Frost heave can occur if there is a supply of moisture being drawn up into surface soil by capillarity from a lower level, causing ice formation which in turn can lift lightly loaded structures. This can lead to differential movement of the foundations if the heave, as is likely, is not uniform. This leads in turn to damage and possible structural failure. When this occurs at or above foundation level, the melted water may not drain down into the underlying strata which may still be frozen. This is common in high latitudes. The soil as a result, puddles and the problem is aggravated. A piled solution is the only remedy, with a pile depth at least four times the diameter or side of the pile penetrating into the frozen soil.

Table 2.2 Frost heave susceptibility

Soil Type	Susceptibility
Silts, chalks, coarse clays	Strongly susceptible
Fine clay (over 40% clay content)	Moderately susceptible
Gravels or coarse sand	Little or no susceptibility

Mining subsidence

When coal and spoil (or other subterranean minerals) are abstracted, natural support is withdrawn and the overlying burden can settle due to gravity. Mining subsidence extends both vertically and horizontally from the width of the excavated section over varying time periods, initially as a wave during mining and later as gravitational settlement. Today, with numerous longwall mining methods to provide case histories, the rate and extent of subsidence can be predicted reasonably accurately.

The extent of subsidence at the surface can be defined by what is known as the angle of draw (see Fig. 2.6). The extent of these angles

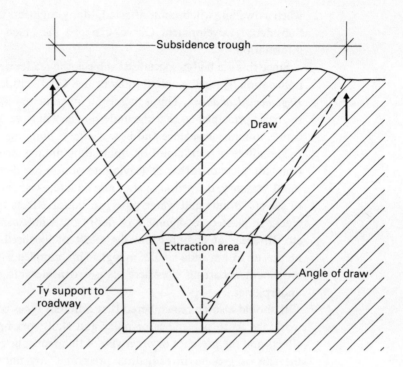

Fig. 2.6 The relationship between angle of draw and subsidence risk

either side of the seam width is governed by the actual seam width and is measured between the vertical of a line joining the edge of the movement area on the surface with the edge of the working seam or roadway excavation. The maximum angle of draw is about 36° but this varies widely and is generally less. In addition the greater the thickness of extraction, the greater the extent of the subsidence.

The maximum settlement immediately above the extraction area is unlikely to exceed 85% of the extraction depth and is dependent on the overlying strata. The subsidence effect lessens with lateral movement at the surface away from the extraction zone and thus a trough of subsidence is created.

Due to the varying movements of the ground at or near the surface both tensile and compressive strains are created. Tensile effects are most pronounced at the outer edge of the trough and compression maximises immediately above the extracted area. Also maximum settlement occurs centrally in the trough with diminishing effects either side. At points of maximum tensile strain, surface fissures tend to appear. It is important to an engineer concerned with subsidence claims to know when peak subsidence has been reached and when movement has ceased, and thus when a dwelling will become affected. Many engineers using the 'Wardell Subsidence Development Curve' can provide much of the necessary information.

Since this is a highly specialised subject, the reader is referred to a 1975 publication by the then NCB (now British Coal) entitled *The Subsidence Engineer's Handbook*. This is merely an introductory attempt to describe the basic principles so that the underpinner will have some idea of what causes mining subsidence and how it affects buildings.

The effect of subsidence on domestic buildings or services can vary from complete disturbance, leaving no option but demolition (i.e. fractured masonry, distortion of steel and concrete framework, and damage to sewers, drains and water pipes) to minor effects of draw including cracks in plaster, re-alignment of doors and windows, floor movements etc. In areas of compression, the effect of localised heave can cause humps up to 1 m wide by 0.25 m high. This, coupled with the differential settlement occurring elsewhere, causes considerable problems for the underpinner.

It should also be remembered that different types of abandoned coal workings, both bell mines or pillar and stall workings, have different effects and the presence or otherwise of these must be established during the site or ground investigation preceding any underpinning works. Another hazard involves capped shafts – these may (and often do) collapse internally causing subsidence and draw effects. Unfortunately, these effects cannot be predicted. Accordingly, site developers or anyone contemplating remedial repairs to subsiding buildings in existing or past mining areas (and this includes salt or ore mines) should investigate carefully the likely effects of possible settlement or subsidence.

Jacking

It is often necessary in areas of continuing mining subsidence to install a jacking system below a strengthened foundation so that as subsidence proceeds the buildings can be jacked back to level. However, this technique would normally only be used in cases where buildings of significant value or status are at risk. It is usual in the case of houses to wait for subsidence to cease and then either repair or demolish and rebuild.

When it is necessary to jack up a building to its former level, it requires exceptional care to avoid further damage. This work should always be carried out by specialist companies. The first requirement is to establish a jacking foundation or base below the existing foundation (see Fig. 2.7). Second, the part to be raised should be free to move without rupturing other parts of the building. Consideration must also be given to the effect

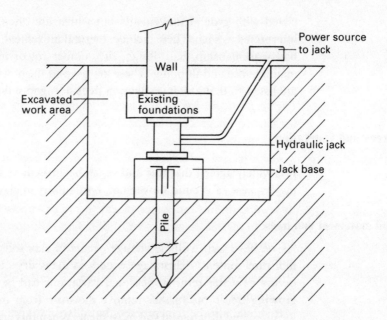

Fig. 2.7 Configuration for a jacking system

on services entering the building. Jacking to restore levels is a last resort before demolition.

Erosion, slope movement, ground decomposition and vibration

In silts and fine sands water movement can transfer fine soil particles away from the ground supporting a building leading to possible subsidence. Ground water movement can be due to: natural causes; seepage from or into fractured drains, sewers or water mains; or the effect of rainwater soakaways from roof drainage along with the direct discharge of rainwater into ground adjoining a building.

A sudden slope movement can subject a building to changed or destructive loading as the ground creeps down to the level of slope.

If building has taken place on made/filled ground, two contingencies are possible: (1) excess compaction takes place; or (2) organics in the fill decompose and release methane and/or carbon dioxide, giving rise to possible explosion or combustion, and ultimately leading to settlement.

Granular type soils can move cohesively under vibration. The source of such vibration can be traffic, adjacent construction work and vibration from nearby manufacturing plants.

There are also several possible causes of damage that are not associated with ground movements but which are often liable to be misinterpreted as such. These include: thermal movement, creep, drying out, bad workmanship, bad design, fire, timber rot or infestation, lack of maintenance and flooding. These are beyond the scope of this book but are listed so that any investigation does not ignore the possibilities.

Trees and vegetation

It is widely agreed that tree and vegetable growth and removal can be a major cause of ground movement, particularly in clayey soils.

Soil movement and trees

As we have seen, volume change occurs in clay soils with variation in moisture content – shrinkage occurs as clays dry out and the reverse occurs as water returns. Seasonal variations can be aggravated when growing tree roots further remove moisture from the soil to produce localised and differential soil movement. When this occurs under or near houses it can produce foundation movement and even structural failure. Buildings on other types of ground other than clay soils are not at risk from this type of movement.

The reverse of soil shrinkage, soil expansion (heave), occurs not only with increases in water availability but also when long-established trees are removed from close to dwellings. Hence heave following the removal of trees is highly likely, particularly in wet weather.

The trees most commonly found near dwellings and associated with building damage are as shown in Table 2.3. Table 2.4 shows the root spreads of some common shrubs.

As a rough guide in the absence of specific information, the height of the tree can be taken as the safe distance from the building.

Assessing

There are four key steps in investigating damage caused by tree growth and removal.

(1) Excavation around the site of damage and settlement should be carried out to ascertain the presence and extent of tree roots, their size, frequency and whether they are active (live).
(2) A sketch should be prepared showing the location, identity and size of trees where roots are causing damage and/or are under or near to foundations. Attention should be paid to the health and sizes of growing trees on account of their increased water demand potential.

Table 2.3 Trees associated with building damage

Tree	Safe distance from dwelling (m)	Type of root
Mature apple	10	Shallow
Ash	17	Deep
Beech	13	Shallow
Birch	9	Shallow
Conifers	5	Deep
Cherries, plum, damsons	10	Shallow to moderately deep
Hawthorn	10	Moderately deep
Horse chestnut	20	Shallow to moderately deep
Holly	3	Moderately deep
Lime	16	Moderately deep
Laburnum	7	Shallow
Lilac	4	Suckers readily and spreads
Poplar	>25	Deep and spreading
Maples and sycamore	16	Deep
Willow	30	Moderately deep and spreading
Yew	5	
Shrubs	Often shallow rooted and less likely to affect building than trees, but can cause local damage by root spread	

There is a relationship between the height of a tree and its safe distance from a building. As a rough guide in the absence of specific information, the height can be taken as the safe distance.

Table 2.4 Likely maximum root spread of some popular garden plants

Berberis	1m
Cotoneaster	2m
Forsythia	7m
Firethorn (Pyracantha)	10m
Hydrangea	1m
Mock Orange	2m
Privet	4m
Climber Roses	3m
Honeysuckle	8m
Ivy	8m
Jasmine	4m

26 *Underpinning*

(3) Tree preservation orders on some trees of large girth may apply and restrict what can be achieved and this aspect should always be investigated.
(4) Evidence of tree removal should be examined by questioning occupiers and neighbours. Any available early aerial photographs or surveys may help.

A considerable amount of advice is available from local parks departments, Kew Gardens and the Building Research Establishment at Watford.

Older houses (i.e. those built before 1950) are likely to be founded on shallow foundations, sometimes as shallow in depth as 0.5 m. Later houses are likely to have deeper foundations – usually between 1 m and 1.5 m. Ideally, foundations should be deep enough to reach a level at which significant volume changes will not occur, and this will often be below 1.5 m. Below such a depth a piled foundation would be indicated, sunk to about 4 m normally. A pile and beam foundation is often used in such cases.

The graph in Fig. 2.8 indicates that 25% of all causes of damage from tree roots occurs within 6 m of the tree and 8% within 10 m. A possible

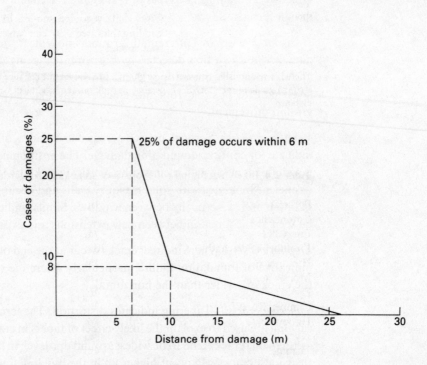

Fig. 2.8 Typical damage incidence caused by trees

safe distance might be set at over 20 m. Again these results constitute only an approximate guide.

Tree removal and pruning

Pruning encourages growth and thus does not reduce moisture absorption. Regular pruning can help to maintain a tree at an acceptable height but this does not guarantee no damage to the house foundation.

Tree removal is an engineering decision where foundations are concerned. Once taken, the tree should be removed in one continuous operation and not piece meal so as not to encourage further growth.

The provision in certain circumstances of a root harness or a root barrier can give protection for perhaps 8–10 years and this method, along with root pruning and height reduction, might even prove permanently effective.

Cracks and cracking as damage indicators

The appearance of cracks can indicate what type of ground movement is affecting a building. The types of cracks and the conclusions that can be drawn are as follows.

Horizontal cracks Differential ground movement and foundation damage are often first detected by horizontal cracking close to ground and damp-proof course (dpc) levels. Horizontal cracks at or about roof level indicate possible roof movement. They may be associated with ground movement.

Vertical cracks Vertical cracks across the facades of premises also indicate (or confirm) foundation damage. The positioning of cracks also is significant along with their widths. Vertical cracking is generally found in ancient buildings with solid masonry walls and occur at the bearings of lintels and/or vertically between windows. Similar indications can also be seen at the junction between new extensions and existing buildings.

Sloping cracks Where indirect cracks occur at the end of walls these are indicative of foundation failure, particularly where the vertical parts of the crack are wider than the horizontal.

Tapering cracks Tapering indicates distortion. The terms hogging and sagging are used to explain the likely effect of tapers in cracks. In sagging (settlement) cracks tend to be widest around dpc level; in hogging (heave) the widest cracks will occur higher up in the building towards roof level (see Fig. 2.9).

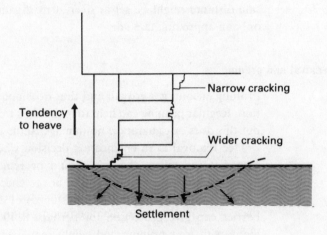

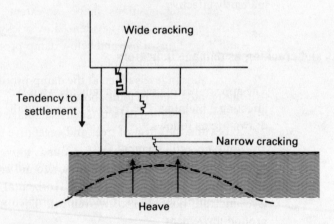

Fig. 2.9 Tapering cracks

Cracks should be viewed only as *indicators* and other evidence should be sought to confirm foundation damage.

Other indicators
Distortion, bulging walls, sloping floors, cracked lintels, sticking doors and windows, and out of plumb walls are all additional confirmation of a preliminary conclusion based on cracks and cracking.

Where cracks coincide internally and externally this is a fairly definite indication of foundation damage, particularly if concentrated on specific access points to the building, when they also help to pinpoint areas of ground movement and likely foundation damage.

The significance of damage and cracking

Initially the building is assessed to ascertain:

(1) whether the damage is structurally dangerous, significant or merely aesthetically unacceptable;
(2) whether the damage has ceased, is continuing or getting worse (progressive);
(3) whether if allowed to continue it will become structurally serious;
(4) what are the likely causes of this damage and are they indicative of foundation movement and ground failure.

In order to arrive at meaningful answers to these questions and to point the way for further, more detailed investigations, the following should be noted:

(1) Cracks (crack positions, shape and extent).
(2) Obvious movements as evidenced by brickwork movement, bulges and distortions above and below damp-proof level, and also cracks in floors and floor slopes etc.
(3) Variations in levels taken at the damp-proof course level or along a designated market joint above this.
(4) External potential defects (such as damage to drains, sewers, water pipes and the effect of trees and vegetative growth or their removal and any infilling etc).
(5) General topography of the area long with any previous history of damage.
(6) Any information of foundation problems in adjacent or nearby property.

This preliminary assessment should provide sufficient data to comment on the significance of the damage and its possible causes and future effect. If ground movement is indicated, a more detailed investigation will be necessary.

Crack and crack profiles are worth further consideration. The severity of these can be classified as follows:

(1) *Aesthetic* – affecting only appearance.
(2) *Utilitarian* – compromising wind and weather tightness and the serviceability of water supply and drainage.
(3) *Structural* – serious danger of further movement leading to possible collapse if not restrained.

The most difficult decisions are associated with cracking in the structural category. These arise when a property is in an advanced state of settlement and the choice falls between repair or demolition. To assist in deciding on the seriousness, BRE Digest 25 (1985) *Assessment of Damage in Low-Rise Building* provides some guidance and should be referred to. It should also be borne in mind that foundation movements can be seasonal, so altering crack profiles throughout the year.

Assessing damage by crack measurement

Crack patterns or profiles

Generally subsidence cracks tend to be diagonal, and tapering from wider to narrow. They tend to start high on the wall or at a door or window and propagate downwards at an angle. To recap:

Slope or incline of cracks Where indirect cracks occur towards the ends of a building they generally indicate foundation failure, particularly where the upper vertical parts tend to be wider than the lower horizontal parts.

Vertical cracking Generally found in older buildings with solid masonry walls, these often appear at the bearings of lintels and/or vertically between windows; similar movement can be seen between new extensions and existing buildings. (See Figs 2.10 and 2.11.)

Tapering cracks The direction of the tapering in cracks invariably suggests some distortion. The terms 'sagging' and 'hogging' are employed to explain the likely effect of tapers in cracks.

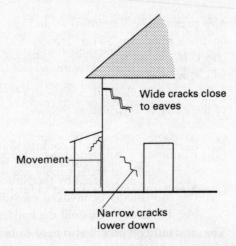

Fig. 2.10 Indication of rotational or differential foundation movement

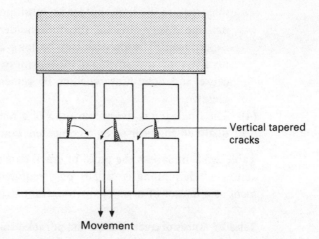

Fig. 2.11 The direction of vertical cracking tapering between openings indicates movement location

Assessment process

When cracks, distortions or gaps appear in a dwelling the occupiers understandably become very concerned. Often they call in the local builder who may or may not be competent to assess the extent of the damage to the structure. It is therefore advisable to seek professional technical advice.

The first step is to establish whether the cause is due to movement in clays. Water ponding in the garden in winter and sticky soil consistency provide useful pointers to clay soils, as do cracks in the ground, particularly flat ones seen in lawns in the summer. Cracking to internal *and* external walls that extends downwards below ground level also indicates clay movement and possible foundation damage as discussed earlier.

The next steps are to measure and record crack widths and distortions so as to assess the likelihood of serious structural damage. If cracks over 5 mm are prevalent and/or wide cracks of 15 mm or more occur, immediate action is indicated. This would then involve the following steps.

(1) Contact the insurers to confirm that subsidence damage is covered by the policy and register a claim. In the case of property under ten years old and covered by the National House Building Guarantee this can usually be invoked provided the building is over 2 years old. Up to two years old the builder is usually still liable.
(2) The mortgagor will also need to be informed where appropriate.
(3) In each case, a loss adjuster will probably inspect the premises and

may require the householder to substantiate the claim. Unless the damage is very obvious, the householder may in turn need to get expert help. The process can be long and drawn out and may involve some monitoring of movement over some time to prove the cause and determine whether movement has ceased or is progressive.

(4) The final step is the preparation of a remedial scheme either by a qualified engineer or by a competent contractor.

Table 2.5 summarises the types of crack damage and possible remedial action. Underpinning is usually only required to arrest further movement as a matter of urgency or to prevent structural collapse.

Table 2.5 Types of crack damage and possible actions

Type of cracking	Action required
Hair cracking below 0.1 mm wide	None
Fine cracking up to 1.0 mm wide	Aesthetic treatment by redecoration
Cracks up to 5.0 mm wide	Fill with plaster or mortars and redecoration
Cracks over 5.0 mm and up to 10 mm wide	Rake out and refill. If external, rake out and repoint including repairs to any defective brickwork.
Cracks 10–25 mm wide	Will require structural repairs often involving the removal of brickwork and its replacement along with rehanging to doors and windows after easing or repair
Cracks over 25 mm wide	Probably will require extensive repairs including demolition in part or whole and rebuilding

Note: The above are guidance limits only. Crack widths are only one facet of the problem and should not be used alone in assessing the damage category.

Crack measurements

The use of glass telltales cemented across cracks is only an indicative method and they can fail to indicate movement due to matrix slip or failure.

The simplest method is repeated measurement with a steel rule and the recording of the distances alongside dates. The location and position of the measuring points should be indelibly marked with waterproof paint. To ascertain relative movement across a crack two measurements should be made at right angles to each other. The use of three round-head brass screws (32 mm No 6s) is suitable. Drill three holes and insert a plastic plug recessed about 6 mm and fill the recess with an epoxy resin before driving home the screws. Position the screws as shown in Fig. 2.12 to form the corners of a right angle triangle with sides of about 75 mm. Measurements should preferably be made with Vernier calipers or in some circumstances a steel rule could be used, measuring from the screws to the limits of the crack.

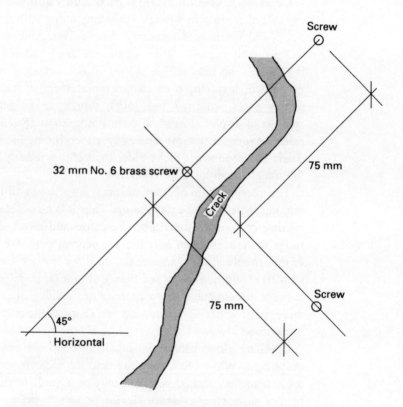

Fig. 2.12 Crack measurement

This measuring or monitoring must extend over at least one year to cover all the seasons and should be resorted to only when there is doubt about:

- whether the cause is ground movement/failure;
- whether the removal of trees has stabilised;
- whether repairs to drains or water mains have had the desired effect.

Monitoring should be carried out at regular intervals until a firm conclusion is reached. It should be noted that in cases of obvious serious structural damage due to ground instability, urgent and prompt remedial works such as underpinning will be necessary and should be carried out straight away without any monitoring. This is a common sense decision.

The frequency of measurements are a matter for judgement. In a very active situation, one to four measurements per month may be necessary. When movement is believed to have ceased once per month for a three month period will provide an acceptable answer.

Generally cracks up to 5 mm maximum width can be considered as capable of remedy by filling or pointing, along with the slight easing of doors and windows if necessary. Above 5 mm and up to about 10 mm will probably require raking out and making good. If no other evidence is apparent this may suffice. Above 10 mm, extensive repair work could be needed, involving even partial replacement of masonry, repairs to service pipes, windows and doors, shoring or bulging walls and the support of displaced joists. If cracks more than 15 mm wide occur, this calls for some structural repairs by underpinning along with easing of doors and windows. Cracks wider than 25 mm usually indicate the need for some demolition and rebuilding.

The interpretation of crack widths is notoriously difficult and set rules relating thereto can be misleading. Thus it is advisable to seek an engineering opinion to determine the cause and effect. The dimensional range for cracks given here are for general guidance only, but above 15 mm should always indicate the need for further investigation with a view to identifying the cause. Sizing of cracks, therefore, is part of any damage assessment. The positioning or profiling of cracks can suggest more, particularly if the direction and shape of the cracking is noted and understood.

Cracking alone cannot be taken as complete evidence of ground movement. While often they can and do indicate ground/foundation movement they should be taken only as a guide to the possibility and further supporting evidence should be sought before reaching a final decision.

Repairs to cracks

Cracks caused by ground movement or failure usually require repair and sealing after the movement has ceased or been arrested by underpinning. Cracking that is less than 5 mm in width does not require any repair other than filling with plaster or mortar (depending on location). Cracks between 5 and 10 mm wide should be sealed with flexible sealants and in the case of external walls by repointing. Cracks wider than 10 mm are more serious and require immediate attention to restore the premises' wind- and weatherproofness and/or to provide support.

Where reinforced concrete is present and cracked, advice on additional treatment of the embedded reinforcement will be needed. The prevention of moisture ingress into the concrete is essential and several proprietary repair systems are available.

Sealants

Today, a wide range of flexible sealants is available. The most popular are simple general purpose mastics and elastomeric sealants composed of sulphide-based polymers, and silicone sealers.

Mortars

External walls are usually repointed in a 1:3 or 1:4 OPC:sand mortar. Cracks can be repaired with the same mortar. However, in the interest of aesthetics it may be necessary to match existing joints with a mortar similar to the original or to colour cement/sand mortars. Similarly the age of the buildings will dictate the type of mortar – older buildings may have been built using a lime mortar, in which case it should be matched by using lime sand or a gauged mortar of lime and sand with a little cement. Table 2.6 provides some guidance on mortar mixes.

Table 2.6 Mortar mixes for crack repair work

Mortar type class-	Hydraulic lime and sand	Portland cement, lime and sand	Masonry cement and sand	Cement and sand plus plasticiser
1	—	1:0.25:3	—	—
2	—	1:0.5:4	—	—
3	—	1:1:5.5	1:4.5	1:5
4	1:2	1:2:8	1:5.5	1:6
5	—	1:2:8 plus plasticiser	—	1:7
6	1:3	1:3:10	1:7	1:8

These are approximately equal strength mixes, giving increased resistance to freezing and improved bond and impermeability.

Concrete repairs

Good practice in concrete repair is to ensure that the repair material matches as closely as possible the characteristics of the original concrete. Repair materials fall into four main categories:

(1) Cement based
(2) Resin based
(3) Polymeric/cement based
(4) Pozzolanic/cement based.

Cement-based materials These are usually mortars or concrete comprising a cement matrix and graded aggregates. The cement used can vary from Portland cements (including sulphate resisting Portland) to high alumina cement with or without chemical admixtures to facilitate waterproofing, accelerated setting, improved workability etc.

Resin based materials There are two main types in use.
(1) *Polyester resins:* These are oil derived resins with good adhesion characteristics and cure rapidly in dry conditions. They set with an exothermic reaction and so can cause thermal movement, thus inducing internal stresses.
(2) *Epoxies:* These are oil derived and are excellent in damp and wet conditions. The setting reaction is again exothermic but not as vigorous as the polyester. Epoxies offer high compressive and tensile strengths and are rapid curing even at low temperatures.

Polymeric modified cement-based material These consist of 'discrete' plastic spheres in an aqueous/cementitious environment. The particles impart increased adhesion to a repair mix, increased flexural strength, reduced shrinkage, improved chemical resistance and impermeability. Among the most popular dispersion materials are styrene butadiene rubber (SBR) and acrylic copolymers, along with a combination of the two.

Pozzolan/cement modified materials Pulverised fly ash (PFA)/fuel ash and micro silica are the most popular pozzolanic materials in use today. A latent hydraulic material (not a pozzolan) also used is ground granulated blast furnace glass. These materials mechanically improve the grading of repair mortars, reduce permeability, and provide good adhesion and low shrinkage properties.

Chapter 3
The survey process and site investigation

The purpose of any investigation is to decide whether ground movement has in fact taken place and what is the most cost-effective remedy. Such an investigation will comprise an initial assessment of damage followed by a more thorough further survey.

Initial assessment of damage

The major concern is obviously *stability*, namely identifying whether there is an unacceptable risk of partial or complete collapse of the building. The next area investigated is *usage*. This is to determine the effect of the movement on work, weather tightness and aesthetics – the effect on the general appearance of the building. Finally knowledge of the site and ground conditions is essential to any underpinning decision.

All underpinning work should be preceded by both a competent site investigation and a ground survey to provide sufficient information to ascertain the cause of the damage and to enable an economic repair scheme to be designed. 'Before underpinning is undertaken, the fullest possible investigation should be carried out to determine whether any underpinning procedure will achieve the object intended' (British Standard Specification 8004).

Evidence of previous underpinning (normally traditional) must be considered because in several cases inadequate works have contributed to the extent and nature of failures. Good advice is available in *Building Research Establishment Digest* Nos 318, 323 and 348 and Code of Practice BS 5930.

To identify the optimum solution, a number of alternative schemes should be examined, not only from the point of view of cost but also in terms of their potential performance, disruption and quality. It should always be borne in mind that there is a difference between value engineering and lowest cost solutions. The latter can lead to technical complications and unnecessary risks.

Given that the period investigations have been properly carried out and the information made available to the contractor is acceptable and satisfactory, an underpinning solution can be produced.

Methods of investigation

The main techniques employed during a site investigation are:

(1) *The desk study* – an examination of published information such as topographical maps, geological maps and records, historical archives, mining records, and photographs.
(2) *The site visit* – to look for evidence of movement, infilled ground, type of foundation, local opinions, trees and signs of growth or removal, and the position of ground water table during summer and winter.
(3) *The soils investigation* – to ascertain soil types, obtain samples, execute in situ soil strength and bearing tests.
(4) *Laboratory tests* – to obtain detailed soil classification and design information.
(5) *Assessment of information and design* – to decide on parameters for design and construction, and to carry out any calculations necessary for settlement, heave and safe bearing capacity.
(6) *During construction inspections* – to validate assessments and design and to alter requirements if indicated.

While the foregoing are all important, the soil/ground investigation is the most essential.

Site and soil investigation

A competent and detailed soil investigation should provide comprehensive data on the following aspects of the project.

(1) Subsoil profiles with information on water table levels for both winter and summer periods if time permits.
(2) Geotechnical properties, including liquid and plastic limits, shrinkability, strengths and descriptions of each soil type encountered.
(3) Variations in ground conditions across the area of the site.

(4) An indication of the probable cause of the damage – settlement, heave, landslip etc.
(5) A triangulation yielding subsoil horizons across the site with depths etc.
(6) The depth, nature and size of the existing foundations.

The most likely cause of any movement may be apparent or deduced, and the likelihood of whether or not movement is likely to be progressive can be considered. However, this cannot be subjective which means long-term monitoring may be required, whereafter a quantitative assessment can be made.

Progressive movement is a problem that is likely to continue and lead to further damage to the structure in the future.

Differential movements arising from variable subsoils across the site or variations in loadings can be difficult to quantify and a comprehensive monitoring regime is often required.

Settlement in granular soils is generally more rapid than clay consolidation and becomes apparent early in the life of a building. It is not usually progressive unless aggravated by external factors such as traffic vibration or changes in the water table. However, silts and clay soils settle by consolidation, a process that is progressive and can affect a structure over a long time, even years. In made-up or filled ground, or in peaty soils, settlements can be dramatic and progressive, with consolidation continuing for a period of years.

Once it has been established that settlement is the likely cause of the damage it is necessary to discover whether or not it has ceased. This can be tested by observation and monitoring as stated earlier, and there are several proprietary monitoring and warning systems available. In most cases underpinning will be required and information as to whether or not settlement is active will be useful in planning the timing of the underpinning and the eventual structural repair works.

Where heave is a major consideration – caused by the expansion of clays affected by trees or tree removal – a different monitoring regime and underpinning solution will be required.

The survey process

The uncertainties present at the design stage of any underpinning solution are both structural and geotechnical. It is at this point also that engineers are at risk. Such risks can be reduced by careful scrutiny of the information available at the time. By this stage, for instance, a full survey of the structure and a competent ground investigation report should have been completed.

Underpinning contractors and specialist engineers understand that adequate site and ground investigations are necessary to produce an economically viable solution. The BSI Code of Practice for Site Investigations (BS 5930, 1981) recommends that the expenditure for investigation should depend on technical considerations such as the value of the premises and the cost of works envisaged to stabilise them.

The ground investigation provides valuable information on soil conditions at the time of the survey, but even these need to be interpreted in the light of possible alterations due to subsequent seasonal and climatic changes, or other constructional operations or movement in the vicinity.

In practice there is often a dearth of reliable information available because of the seemingly prohibitive cost of performing a comprehensive site investigation, and the tendency of clients and specifiers to rely on the expertise of the contractor. Their reluctance to invest in adequate soil surveys means the contractors either carry out their own limited investigation or even rely solely on their experience in order to provide the required competitive margin. This lack of information militates against a client receiving any guarantees. Money spent on acquiring adequate ground data will also lead to better schemes, greater cost effectiveness and probably enable guarantees to be given.

The success of any underpinning scheme depends on a sound underpinning design and proper execution. This is not possible without a thorough understanding of the physical nature of the structure of the ground, and of the site. Hence, both site and ground investigations are essential prerequisites to good design.

The site assessment

This is a practical but initial examination aimed at determining whether the damage is significant; whether it is progressive or static; how serious it is likely to become if not dealt with; and the most likely cause. It also aims to provide background and historical information on the site. It takes the form of a site inspection, desk research, a topographical study, examination of photographs, a 'dilapidations' survey to investigate damage due to wear and tear, and talking to the occupiers and local residents. It is not normally an expensive process.

Desk study

A study of old Ordnance Survey maps, geological maps, local authority and archaeological society archives will minimize the surprises which usually occur during construction operations. Such a study will often provide evidence of tipping, wells, watercourses and so on under and around the premises. Useful clues can be obtained from an examination of nearby road and place names. For example, Wet Lane, Claypit Lane, Springbank and Quarry Road could be names that prompt further investigation.

If there is work going on nearby, it will often be rewarding to visit the site and make local enquiries. Local Building Control Offices are usually very useful sources of information on their own areas. An investigation of the statutory and public utility bodies' services should be made and verified before embarking on any subterranean works.

Underground services

These should be located and checked, particularly the drainage, sewage and water supply services. Leaking drains and water supply pipes are often responsible for the kind of local subsoil softening that can adversely affect nearby foundations.

Points to note

An initial site assessment should include and report the following things.

- The position, size and estimated age (old or new) of cracks should be recorded, preferably with annotated and dimensioned sketches and photographs.
- Movements of brickwork, masonry, lintels and floors should be similarly treated.
- Displacement of doors and windows, 'refitted' doors, broken glass and arched thresholds should be noted.
- An inspection of the site following rain is often useful.
- Adjacent buildings should be inspected for signs of damage.
- Checks should be made for slips or irregularities in kerblines and footpaths.
- The vegetation on the site should be mapped, with particular reference to trees that have been removed or cut back.

An analysis of the data from this assessment will facilitate a meaningful interpretation of the next stage of the investigation.

Ground investigation

The costs of ground investigation for an underpinning project could be between 2 and 12.5% of the cost of the works. The normal expenditure is around 5% of the total. Although it may be a priority to minimise costs, it should be borne in mind that the quality of the investigation obviously affects the reliability of the data obtained. And this data is a vital element in any successful underpinning project.

Guidance can be obtained from Codes of Practice and various standards, and the techniques of ground surveys are well known to most engineers. Further information is available in the following publications:

- BS 5930: 1981 *Code of Practice for Site Investigations for Low Rise Buildings*. BSI, London.
- BRE Digest 275, *Fill – Part 2: Site investigation, ground improvement and foundation design*, Building Research Establishment, Watford.
- Clayton, C.R.I., Matthews, M.C. & Simons, N.E. (1995) *Site Investigation*, 2nd edn. Blackwell Science, Oxford.

A ground investigation is best carried out by a specialist contractor. In effect, their report is the first step in the design effort. The objective of the ground investigation is to determine the sequence, thicknesses and types of the various strata along with ground water levels to a depth greater than the foundation levels, and preferably to locate competent lower ground. Specifically the report on the investigation should:

- provide general descriptions and thicknesses of the soil at various depths;
- provide bearing capacities where appropriate (or opinions thereon);
- provide a detailed soils profile, identifying any variations in strata across the site;
- determine ground water levels;
- give data on harmful or deleterious chemicals present, particularly sulphates; and
- provide any other potentially useful information.

In the case of domestic property it is normally adequate to provide a soils profile down to a depth of 5 m. The work is carried out by digging trial pits in appropriate positions. A typical profile is shown in Figs 3.1 and 3.2. It is usual to preface the report with a general descriptions and summary (see Tables 3.1 and 3.2).

Trial pits can be hand dug or machine excavated down to the required depth. They provide adequate information down to about 5 m. Below this depth, boreholes (usually 150 mm in diameter) offer an economic

Profile

Sampling/In situ test			Description of strata	Strata			
Depth (m)	In situ tests	Sampling		Legend	Reduced level	Depth	Thickness
			Dark grey brown fine sandy silt abundant roots (TOPSOIL)			4	4
						5	2
			Dark grey silty CLAY with occasional rootlets (TOPSOIL)			7	8
			Light grey to green grey fine sandy CLAY with occasional fine angular gravel and rootlets (BRACKLESHAM BEDS)			1.2	3
1.3	▽	5	Light brown silty well graded SAND and angular to sub-rounded GRAVEL (BRACKLESHAM BEDS)				1.1
	▽	7					
	▽	9	Medium dense grey green silty fine SAND with occasional decomposed roots and rootlets. Very strong smell of hydrogen sulphate (BRACKLESHAM BEDS)			2.5	
	▽	10					4
	▽	14					
			Medium dense grey silty fine SAND (BRACKLESHAM BEDS)			2.7	

Ground water record				Key	Notes
Strike(m)	Rate	Time		■ Undisturbed	
				☐ Core	
				▽ Earth probe 150 mm	

Fig. 3.1 Typical soil profile from the pit positions shown in Fig. 3.2

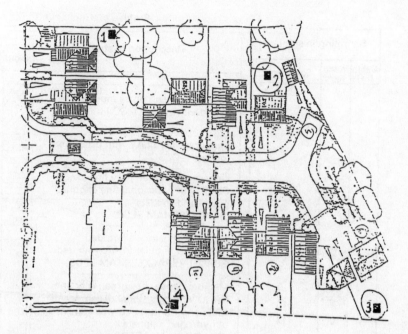

Fig. 3.2 Block location plan

Table 3.1 Typical description of ground/soil conditions

Strata	Depth (m)	Thickness (m)
Dark grey brown sandy silt (TOPSOIL)	0.0–0.4	0.4
Brown to green grey sandy silty CLAY (BRACKLESHAM BEDS)	0.4–0.8	0.4
Medium dense grey geen silty fine SAND (BRACKLESHAM BEDS)	0.8–2.3	1.5
Medium dense grey to orange brown silty fine SAND (BRACKLESHAM BEDS)	2.3–3.0+	0.7

method of determining conditions at greater depths. Mini soil surveys, with a diameter of 65 mm, are also used for such assessments.

Soil tests

There are two types of samples – disturbed and undisturbed.

(1) *Disturbed samples* can be obtained from all types of soil and, if tested promptly, offer the opportunity to determine moisture contents and identify constituents. Disturbance resulting from the

Table 3.2 Typical description of surface/ground conditions

Surface Conditions

The site comprised a flat level area of both grassed and wooded land. Crossing the site there were a number of shallow ditches. It was reported by the owner of the property that there was a rubble infilled swimming pool adjacent to Trial Pit 4.

Ground Conditions

Ground conditions were generally as anticipated from the desk study. Generalised ground conditions are presented in Table 3.1.

For more detailed information reference should be made to the Trial Pit Logs.

sampling process means that measurements of mechanical properties will not be reliable.

(2) *Undisturbed samples* can be taken, with great care, only from cohesive soils. If acquired properly these samples will give test results that provide an accurate idea of how the soil will behave in situ.

The soil tests that can be carried out include both laboratory and field tests.

Laboratory tests

- Identification and classification of soil types
- Shear and compressive tests
- Moisture contents
- Densities
- Particle size analyses
- Consolidation tests

Field tests

- Vane test for shear strength
- Standard penetration tests
- Bearing tests

The most useful to the underpinner is the standard penetration test (SPT). In the UK this is currently used primarily in non-cohesive soils (such as sands and sandy gravels) and consists of driving a probe of specified dimensions into the soil in a standard manner using a weight dropped from a given height. The probe is driven down 450 mm and the blows are counted for each 76 mm (3 in) of penetration. The penetration resistance (N) is the number of blows required to drive the probe for the last 300 mm. The standard penetration N value can give an indication of soil stiffnesses and load-bearing capacities which are crucial in designing foundations and underpinning solutions. It is a fairly cost-effective test and is universally used.

SPT techniques vary across the world but there is currently a great deal of effort being put into devising standard methods of interpreting the results and perfecting the test. For more detailed descriptions, see BS 5930 (1981) *Code of Practice for Site Investigations* and BS 1377: Part 9 (1990) *British Standard Methods of Test for Soils for Civil Engineering Purposes, Part 9: In Situ Tests* and the appropriate ASTM (American Society for Testing and Materials) document ASTM D1586–84 (re-approved 1992) *Standard Test Method for Penetration Test and Split-Barrel Sampling of Soils*.

A useful table of relative densities (from US ASTM D2049–69) is shown in Table 3.3.

Table 3.3 Assessment of states of compaction from relative densities

Relative density	State of compaction
0–15	Very loose
15–35	Loose
35–65	Medium
65–85	Compact
85–100	Very compact

The subject matter covering the geophysical properties of soils is both extensive and complex and beyond the scope of this book. It is always wise to seek the advice of a qualified specialist. Reference can be made to G.A. Leonards' *Foundation Engineering* (McGraw-Hill, New York, 1962) or to *Foundation Engineering*, edited by G. Pilot (Presse de l'Ecole Nationale des Ponts et Chaussees, Paris, 1982) or to any good recent foundations manual.

Once the background research and investigation has been done, it is possible to move on to the selection and design of the underpinning system required to remedy the problems encountered.

Whether or not to underpin

Generally, these decisions are subjective but they can be improved by using to the full the techniques already outlined. No work should be carried out until the cause of movement has been established.

Then, options must be examined:

(1) Repairing and strengthening of the structure itself
(2) Tree removals or root pruning/drainage
(3) Underpinning
(4) Final building repair.

A suggested approach is shown by the flow chart in Fig. 3.3.

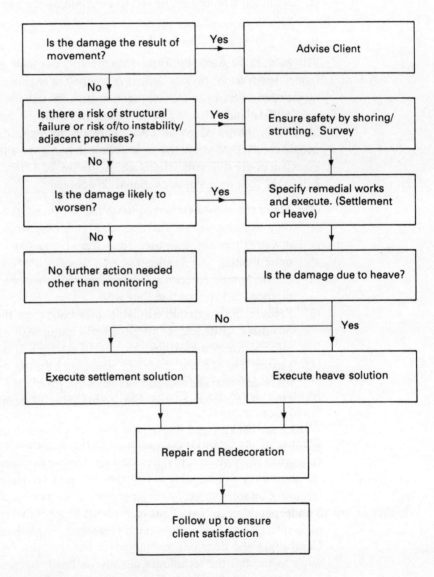

Fig. 3.3 A decision-making process for underpinners

What will a site investigation achieve?

There is no guarantee of success with a site investigation. How many problems are detected will be a function of:

- how technically competent is the person carrying out the work;
- how experienced they are, particularly with the local ground conditions and the type of remedial work envisaged;
- how much freedom is allowed in determining the methods of investigation; and
- how much money is allocated.

The balance between sufficient reduction of risk and the amount of money spent is not easy to define. A competent engineer or contractor might spend very little on investigation because the ground conditions are familiar and relatively predictable, but could be just as likely to identify problems which ideally require further expenditure. No two people are likely to adopt the same approach to a given site because their experiences are different. Therefore, to get the best from a site investigation the following methods should be adopted.

(1) Make the person employed to do the site investigation aware that they are responsible for designing and executing a satisfactory investigation for purposes that you define (e.g. the proposed underpinning of a dwelling).
(2) Allow complete freedom (within financial constraints) to use any methods of investigation they wish.
(3) Provide the maximum available information on the dimensions, positions, loads, etc. of the structures along with any other local information you have.
(4) Ensure that the site is visited before and during operations and make any changes necessary as the work proceeds.
(5) Discuss the findings and seek clarification of points not understood.

Because of the financial constraints that the householder or insurance company may wish to apply, it is unlikely that cautious persons carrying out the survey will be liable in contract, but they certainly will be liable for using reasonable skill and judgement in carrying out their duties. This does not mean that all problems must be identified since the test is of comparison with what an ordinary reasonable competent practitioner could achieve.

Conclusions

(1) Site investigation must be relevant to the site, the ground and the dwelling. It is specifically a specialist field, and requires skilled specialists for successful results.
(2) Site investigations for shallow foundations require special attention to desk studies, local knowledge, and site visits both before any ground investigation and during active construction.
(3) The value of site investigation can be increased by giving a competent, qualified and experienced geotechnical engineer the responsibility for the investigation.
(4) A site investigation will not always detect all the adverse conditions which may affect foundations, but it does and can minimise risk.

Chapter 4
Traditional underpinning

Although it is often expensive, labour intensive and disruptive, traditional underpinning is still the most common method of tackling ground movement problems in low rise buildings, especially housing. The majority of such work is carried out by local builders. Specialist contractors prefer piled solutions as being more cost effective but these are probably beyond the scope of most contracting builders.

Mass concrete systems

Traditional underpinning involves stabilising existing walls by digging out the soil beneath the foundations to a depth where firm, competent strata exist and replacing it with mass or reinforced concrete. The depth of the excavation below the foundation does not usually exceed 1.5 m and the concrete infill should have a compressive strength of 20–30 N/mm^2 at 28 days. The idea is to transfer the load on the existing foundation to deeper and more suitable load-bearing strata, so stabilising shallow foundations. Hence the bearing capacity of the ground needs to be assessed. This should preferably be done by load testing but can also be gauged by experience of previous work on the site. It is always worth seeking professional advice if time and budget permit (see Chapter 3). Table 4.1 gives an indication of the bearing capacities of some common soil types. More general characteristics are given in Table 4.2.

The work is carried out in 'discrete' or alternate bays – generally up to 1.4 m in length and up to 0.6 m wide with depths of 1.5 m or more below ground level – so as to maintain sufficient support as the work proceeds. No more than 20% of the total wall length should be left unsupported and the unsupported length at any one time in each bay should not exceed 1.4 m. The bays are then joined together to form a continuous beam as shown in Fig. 4.1. (See also Fig. 4.2.)

It is most important not to over-excavate and thereby expose the

Table 4.1 Approximate bearing capacity of some common soils

Soil	Bearing strength (kN/m^2)
Compact gravel, dense coarse clay	> 600
Stiff clay	150–300 (220)
Firm clay or sandy clay	70–150 (110)
Loose gravel, loose sand or loose silty sand, changing sand	100–180 (140)
Soft silt, clay/silty clay	75
Solid chalk	600
Plastic chalk, very soft soil and clay	Need specialist assessment

Table 4.2 General soil characteristics

Gravels and sands, coarse grained or granular, cohesiveless soils	Silts and clays, fine grained cohesive soils
Low proportion of voids between particles	High proportion of voids between particles
Slightly compressible; compression under load is rapid	Highly compressible; compression is slow
Permeable	Virtually impermeable
Little cohesion	Good cohesion
Little variation in volume with changes in moisture content	Considerable change in volume with change in moisture content

entire length of the flank of the footings to be underpinned. Removing lateral support in this way will, in some soils, permit rotation or side movement to occur and this is the most likely cause of building collapse or additional damage. Obviously underpinning should always aim to correct damage rather than add to it.

The system would normally be used because of uneven settlement, increased loadings or the loss of support from adjacent ground. It is not always necessary to carry out the work to form a continuous beam; a variation often referred to as 'hit and miss' can be used when the original foundation is in good condition or perhaps reinforced. This involves forming bays as already described, but leaving gaps between the 'concrete columns' (i.e. omitting some bays).

Where part of a building is on competent ground but another part has subsided, only the defective area is underpinned (partial underpinning). This requires very careful consideration because of the risk of differential movements. In the case of partial underpinning, the work should extend at least 1.5 m beyond the area underpinned, and it should always be

52 Underpinning

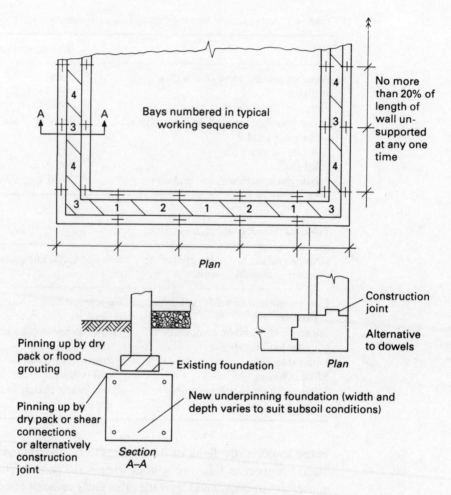

Fig. 4.1 The traditional underpinning system

returned a similar distance around any corners – see Fig. 4.3. This allows the building load to be spread through the fabric of the building and across the transition distance.

Advantages

The advantages of the traditional method are:

- The engineering principles are simple and easily understood.
- It offers a low-cost solution at shallow depth – less than 1.5 m.
- The work can be undertaken from one side of a wall, so occupants do not necessarily need to be relocated.

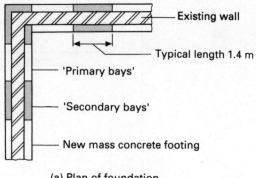

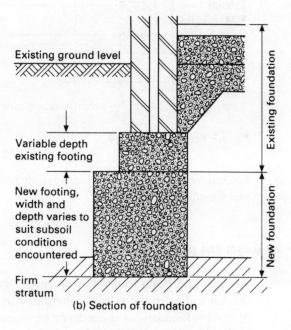

Fig. 4.2 An alternative version of the traditional system

- It is suitable for most foundation loads.
- The system is suited to supporting construction where deepening is required in competent subsoils.
- Work can be carried out in areas of difficult and restricted access.
- It is especially suitable for the formation of cellars and basements beneath existing structures.
- The work can be carried out by general builders.

As the depths of component strata increase, the advantages of the mass concrete system are nullified. This has given rise to an alternative

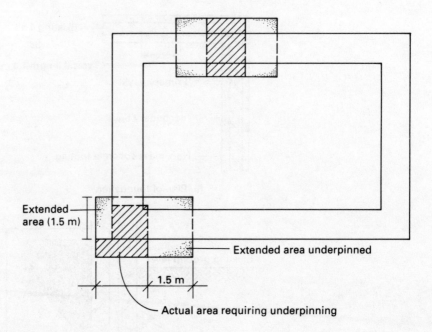

Fig. 4.3 Partial underpinning

method, namely the pad and beam system (often referred to as beam and pier foundation).

Pad and beam (beam and pier) systems

The pad and beam foundation technique was developed to provide a quicker and more cost-effective alternative to mass concrete stabilisation of shallow foundations at depths down to approximately 4 m. In essence it involves the use of a reinforced concrete beam supported by mass concrete piers extending down to a competent soil layer and is hence still considered to be a form of traditional underpinning.

Beyond a depth of 1.5 m, mass concrete underpinning loses its cost advantage, becoming increasingly labour intensive and time consuming. Between 1.5 and 4 m, and where further ground movements are anticipated, the pad and beam system is the most viable underpinning solution. The method is illustrated in Fig. 4.4. It involves constructing a reinforced concrete beam within the brickwork. It is constructed in short lengths (2–3 m) by breaking out short lengths or holes in the wall to be supported, and propping them with steel, concrete or brick stools which are sacrificial, and subsequently cast into the concrete beam.

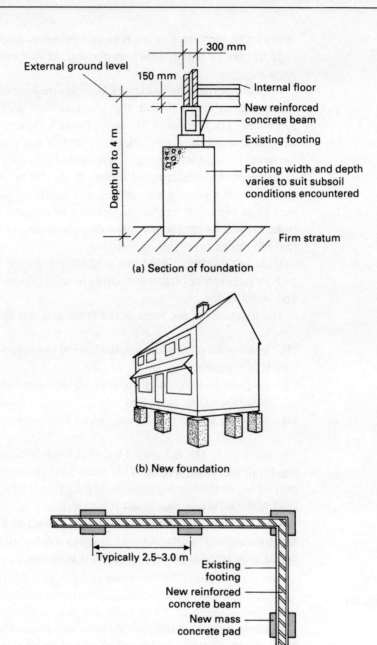

Fig. 4.4 Pad and beam foundation

When sections of walling are removed before underpinning, temporary support may be required such as pinning the wall or providing external support.

Once supported, the brickwork between the supports is removed, the reinforcement inserted and the concrete cast within any necessary formwork. The brickwork above the beam is shimmed or packed where necessary, using slips, new brickwork and/or dry packed mortar. After the beam is cured and has achieved its strength, concrete piers are then constructed at predetermined centres along the beam to a depth sufficient to reach a stratum with the required bearing capacity. In the case of very large piers, the weight and cost can be reduced by the use of void formers such as cardboard tubes or polystyrene block-outs, provided that the design takes this into account.

If the suitable bearing stratum is at a depth greater than, say, 4 m, the pads or piers can be further supported by mini piles to form a type of pile and beam system.

The functions of the beam in the beam and pier system are:

(1) to span the piers and carry the load of the superstructure for which it is designed;
(2) to provide temporary support to the structure during the construction of the piers; and
(3) to tie the structure together and so improve its overall rigidity.

The piers carry the designed load plus loads generated by the beams together with their weight. This total load is carried by the area of ground where the bearing capacity of the strata is adequate to sustain the load without further significant settlement.

The total load of X tonnes bears on the ground of Y tonnes per square metre capacity. The area of the pier in plan is calculated as $X/Y\,(\text{m}^2)$ after taking into account any relevant factor of safety.

Stools detail

The stools used as sacrificial supports can be purpose made concrete blocks of an appropriate size, cut lengths of rolled steel joints or purpose made cast iron units. In some cases it may be adequate to use layers of engineering bricks at set intervals.

Normally the width of the stools should be 50 mm less than the wall they support and they should be placed at the centre of the width of the wall at approximately 1.0 m intervals (i.e. at least two per section).

Reinforcement

The beam should be designed to span between piles and contain sufficient reinforcement in the form of horizontal rebars top and bottom secured by hoops. All steel should have at least 40 mm of concrete cover all round. (BS 8110 gives details of the requirements.)

Void formers and heave precautions

If there is a likelihood of heave occurring, a void potential should be formed around the top 1 m of the pad and pier by use of a fibre board which would allow slip or movement by squeezing. The sides of the beam itself should be backfilled with polystyrene of a density not exceeding 11 kg/m^3 or with suitable compressible fibrous material. A void should also be left beneath the suspended floor. An alternative is to use loose backfill.

Sulphates, chlorides and aggressive agents

The presence of the above in soils or gravels can be detrimental to concrete and necessitate higher quality concrete and/or the use of special cements. Only adequate chemical testing will provide sufficient data for appropriate action.

Generally when the total sulphate content (expressed as sulphur trioxide (SO_3) is below 0.2% in soils or below 0.3% in ground waters, ordinary Portland cement (OPC) content in excess of 250 kg/m^3 and a maximum water to cement ratio of 0.5 provides adequate safeguard. Where the sulphate content increases to 0.5% in soils and above 0.3% in ground water, the minimum OPC content needs to be 330 kg/m3 and the water content ratio should again be 0.5.

Above these limits for sulphate control it is advisable to use sulphate resisting Portland cement or a pozzolonic cement. Again, seeking professional advice is normally a sensible precaution. The current *Building Research Establishment Digest* on the subject contains further information.

Advantages

The advantages of the pad and beam/beam and pier system are:

- It is cost effective at depths down to 4 m below ground level and also with heavy foundation loads.

- It can be carried out from one side of the wall so occupants do not necessarily need to be relocated.
- The work can be carried out in areas of difficult and restricted access.
- The system is suitable for supporting bridges made of stone.
- The system is effective in clay soils that are susceptible to heave.
- The work can be undertaken by local builders given an acceptable beam design.

Chapter 5
Piling in underpinning

Experience in mainstream piling has inevitably led to the use of piles in underpinning systems. In low rise buildings, restrictions of access, height and space have been overcome by developing small specialised compact plant and equipment offering quiet and, if necessary, virtually vibration-free installation of the piles, even inside the premises.

The systems available can be grouped conveniently into three categories:

(1) Foundation stabilisation from one side of the wall only, preferably externally, though not essentially. Systems include simple angle piles, cantilever piled beam, pile and knuckle and traditional mass concrete.
(2) Foundation stabilisation from both sides of a wall, including dual angle piles and pile and beam systems.
(3) Specialist systems, such as pali radice, pin piling, pressure grouting, jack-down systems and heave systems.

Conventional piles can be either displacement or non-displacement types. The former are thrust or driven and the latter generally are excavated or drilled. The piles used can be steel tube, precast concrete or in situ concrete. A variety of shapes and sizes is in use, but square or round concrete and round tubes predominate. Sizes range from micro or mini to large piles. A useful definition of sizes is:

- *Small* 300–600 mm, side or diameter
- *Mini* 65–300 mm, side or diameter.

Some engineers split the mini classification into mini (150–300 mm) and micro (65–150 mm). In domestic underpinning the 65–300 mm mini is the preferred definition.

Piling techniques

Angle piling (raked piling)

This type of system is used to stabilise all types of concrete strip foundations in low rise buildings and generally uses steel tube mini piles. (See Figs 5.1 and 5.2.) The piles are normally set close together at, say, 1–1.5 m centres. Because of the short distances between piles, the load carried on each pile is small, allowing the smaller diameter, mini piles to be used.

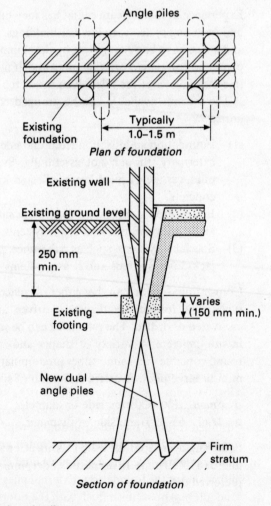

Fig. 5.1 Dual angle piles

Piling in underpinning 61

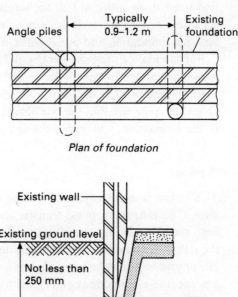

Fig. 5.2 Single angle pile

The existing foundation is pre-drilled or cored at an angle using air flushed rotary percussive equipment or other suitable coring means. Permanently cased steel or plastic driven or augered piles are then installed through the pre-drilled hole, with the casing terminating at the underside of the reinforcement included up through the foundation.

The piles may be single or double, and are generally placed alternately either side of the foundation. External piles finish just below ground level and internal ones finish flush with the floor slab or over site concrete but below the topping so as to provide an accurate floor finish.

The system relies on the existing foundation having sufficient depth and strength to withstand all the stress imposed by the reaction of the structure on the piles. It is important that the bond and adhesion between foundation and pile is sufficient to withstand these forces.

For foundations/footing depths of 250 mm or thereabouts a 10° rake is usually enough to ensure the pile passes through the actual centre of the foundations. The drill angle should never exceed 15° and the aim should be to ensure that the pile always passes through the middle third of the foundation, and within the footprint of the actual wall.

Pin piling (floor slab piling)

This system is used to upgrade existing domestic or light industrial floor slabs. The pins (mini piles) transfer the load directly through the slab onto the new adequate bearing stratum (see Fig. 5.3). The adequacy of the slab to span between the pile positions governs the distance apart of pile or grid and must be assessed. It is usual to place them approximately 1 m apart in all directions and in the form of a uniform grid pattern.

This system uses mini piles normally about 1.5 m in length. The technique is to drill or core through the concrete slabs or raft and insert mini piles of either 65 mm or 90 mm diameter to the predetermined grid pattern with the head of each pile set into the floor slab by grouting up or concreting the pile head. The grid pattern will of course depend on the thickness, strength and type of floor slab being piled.

The piles when installed will fix a slab in position even when the ground immediately below is incapable of supporting the underside of the slab. In this latter case it is advisable to grout up any set voids before piling. Grouts can also be used to 'jack' the slabs back to line and level before piling. Where machinery is installed (or is to be installed) on the slab, additional piles can be located beneath the machines so as to transfer the load directly to a suitable bearing stratum and thus directly support them on piles. These individual piles may be driven to a deeper depth than the main pin piles or, in the case of heavy machines, could be of a larger size.

Needle piling (pile and beam)

This technique involves the installation of mini or other piles to carry reinforced concrete or steel needle beams running through the walls (see Fig. 5.4). The load is supported at close centres of about 1 m spacing, either by directly spanning between opposite piles either side of the wall where practical, or by cantilevering the needle over two external piles.

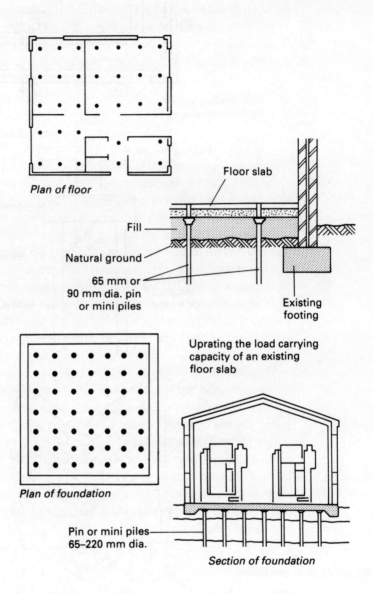

Fig. 5.3 Pin piling

64 *Underpinning*

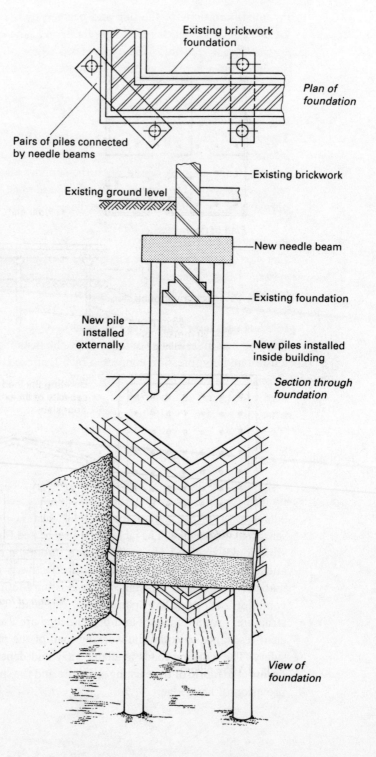

Fig. 5.4 Needle piling (pile and beam)

Patented variations of the pile and beam system are also used. These include pile and knuckle beam, lateral pile and cantilever, and are used in domestic and light industrial applications where only external access is possible, and where the existing foundation is generally in good condition.

Pile and knuckle beam
The pile and knuckle (Fig. 5.5) uses one vertical and one raked mini pile, interconnected through the wall by a reinforced concrete needle beam. Loading capacities are based upon the support afforded by the underlying ground. The centroid of the applied load must be between the mini piles and at the underside of the needle beam.

Pier system knuckle
The pier system knuckle is an adaptation of the pile and knuckle beam system. It uses a single concrete pile reinforced and connected to the foundation by a tee beam (see Fig. 5.6).

Pile and cantilever (cantilever ring beam)
This method stabilises a foundation by the installation of mini piles in tension and compression connected by a reinforced concrete ring beam incorporating concrete needles to support the wall (see Fig. 5.7). It is used where adequate bearing strata are at depths in excess of 1.5 m and where longitudinal stability is also required.

Jack-down piles

Jack-down piling is used when other forms of underpinning are uneconomical because of extreme depths of suitable load-bearing strata, or where other piling systems could cause further problems due to the sensitivity of the area. The jack-down system (see Fig. 5.8) is a silent and vibrationless method of upgrading and stabilising existing foundations and column bases, and is highly cost effective, with every pile automatically tested as it is installed. The existing footings must be in good condition since they will span the pile cap head. The dead load of the structure, ground anchors and/or kentledge are used to generate sufficient resistance to equal the working load of the pile plus its factor of safety. The working loads available vary and depend on: ground conditions, the fabric of the existing structure and the space available for the provision of kentledge.

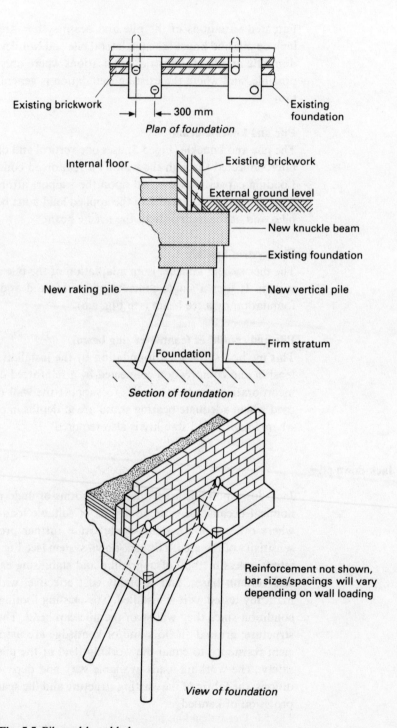

Fig. 5.5 Pile and knuckle beam

Piling in underpinning 67

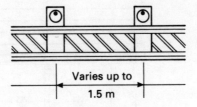
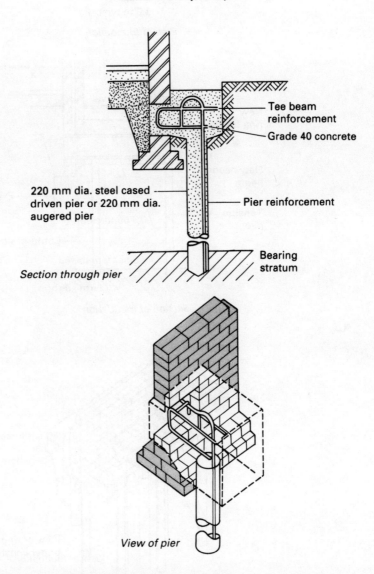

Fig. 5.6 Pier system knuckle

68 *Underpinning*

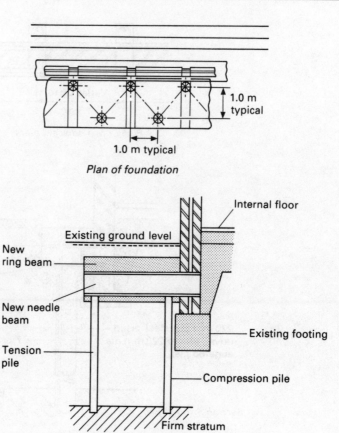

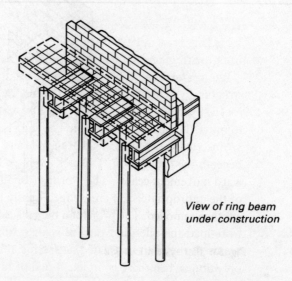

Fig. 5.7 The pile and cantilever (cantilever ring beam) system

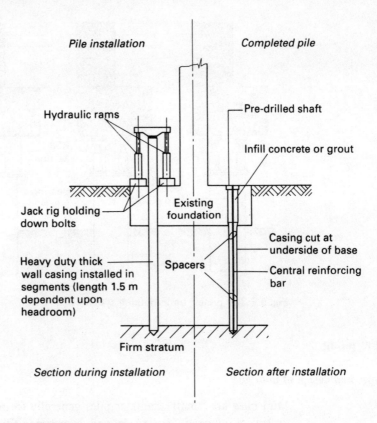

Fig. 5.8 Jack-down piles

Pretest piles

The 'pretest' pile system (Spencer White and Prentis in the USA) uses 1.2 m lengths of open ended steel tube jacked into place using jack-down from the foundation being underpinned (see Fig. 5.9).

A pit is excavated below the foundation (which must be competent structurally in itself to withstand the jacking) to provide a working space for the hydraulic jacks (400 kN capacity) which apply the necessary force to move the pile down to a safe bearing. Then a pair of hydraulic jacks are inserted between the head of the pile and the underside of a steel plate beneath the foundation. The jacks apply a load on the pile until downward movement ceases. Then, props (usually cut from rolled steel joists of the appropriate size) are wedged tightly between the jacks. The jacks are then removed. The props can be replaced by jacks again in the future to re-load the piles and then re-wedge. Alternatively, if permanence is required, they can be treated as sacrificial and concreted in, but this is only done when there is no likelihood of future movement.

70 *Underpinning*

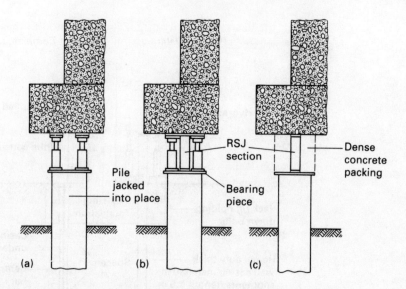

Fig. 5.9 The pretest underpinning system

Mini piling

Types and design of mini pile

Mini piles are 'small' diameter piles generally formed of hollow steel, driven or augered into place and sometimes filled with a suitable cementitious grout or a flowable concrete. Nominal reinforcement is provided usually by a single small diameter mild steel bar if required. Plastic tube filled with concrete may also be used. The base of the pile is closed, in the case of steel tube by a cruciform closure made under pressure using a mandrel; in the case of plastic by a purpose made end piece.

Normally, mini piles are driven to a 'set' (a point at which a desired penetration resistance is reached) or augered down to a stable base (depending on conditions). Care is needed in the case of 'made or filled' ground to ensure that the set is not founded on a buried obstruction, even though, in practice, because of the close spacing of these economical piles, this may not constitute a serious risk. However, it is always advantageous to have the fullest possible knowledge of the ground conditions and previous history of the site prior to starting work. (See Chapter 3 for a description of site and ground investigation.) Mini pile systems should always be designed based on known driving criteria.

Any pile derives its ability to support loads, up to the bearing capacity of the ground in which it is founded, from:

(1) the bearing capacity of its base in relation to the ground; and
(2) frictional support along its shaft.

In firm clays, friction plays a dominant part in determining the ultimate bearing capacity, whereas in soft clays or loose granular material there is minimal frictional support. In domestic underpinning it is usual to consider only the bearing capacity and apply a safety factor in the design.

When hollow steel tubes are being used, either top or bottom driven, an empirical set is said to be achieved when 10 blows of the driving force cause a penetration of less than 25 mm. This is invariably adequate in the case of domestic underpinning. In the case of a vibratory hammer, a penetration of 10 mm in 10 seconds applied over a penetration distance of 100 mm is an acceptable standard. Obviously, the amount of energy being expended will affect the results. If greater accuracy is deemed necessary, there are formulae used in mainstream piling that can be used to define the required penetration resistance much more precisely.

A period of rest should be required after an initial set has been achieved and, following such a break, a re-drive should be undertaken to ensure that the required set still pertains. This is necessary in some types of soils because water between the solid particles can have a lubricating effect that may reduce the bearing capacity over time.

In low rise properties, underpinning piles are generally in compression. In some systems, piles are used as anchors (e.g. the cantilever system). Where angle piles are used or the loading in vertical piles is eccentrically applied the bending moment can be reduced but experience shows that this is not critical, as the load will realign itself to act through the pile centre.

To cater for buckling it is generally thought the lengths of mini piles should be restricted to about 70 times their internal diameter. However, over the past 20 years, experience has shown that this criterion can be exceeded by a factor 2, depending on ground conditions. In driving many kilometres of small diameter piles, no evidence of significant buckling has been observed by the present authors.

Design of underpinning solutions using piles is covered in more detail in Chapter 6.

Mini pile production

Mini piles up to 150 mm diameter are usually factory produced from lightweight steel casings at between 1.8 and 2.6 mm wall thicknesses. They are produced in various lengths and can be joined together by a male and female type socket produced by 'swaging' one end. The bottom

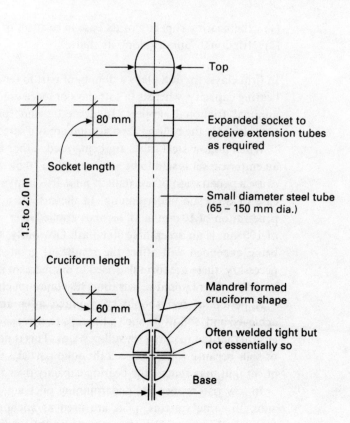

Fig. 5.10 Typical mini pile

of the first section to be driven is closed by a cruciform type end as shown in Fig. 5.10.

The socket is generally about 80 mm long, whereas the cruciform end length varies from 60 mm for the 65 mm tube to 180 mm for a 150 mm diameter tube.

When piles longer than about 2 m are required, they are usually sectionalised and the extension pieces joined by a socket push-fit joint of the extension into an enlargement of the longer piece. These are sometimes spot welded into position, depending on conditions or specifications.

Piles above 300 mm diameter are designated 'small' rather than 'mini' and are made from thicker steel casings of about 5–6 mm wall thicknesses. These again have a cruciform end and similar socket joint but at this size they will be fitted with a welded collar for strength and continuity. Mini piles are usually top driven through a dolly cap, whereas small piles may be either top or bottom driven.

Mini post grouted piles

These are in situ mini piles formed from a small diameter drilled hole which is pressure grouted and reinforced with a full length reinforcing bar. These piles derive support mainly from the soil by skin friction as the pressure grouting enhances the bonding between soil and grout, thus increasing their load carrying capacity. The method of construction is as follows.

(1) Drill a hole roughly 127 mm in diameter (5 inches) down to the required depth and install a full length casing.
(2) Insert the reinforcing bar (rebar), located centrally by means of spacers, and fit grouting tube at same time.
(3) Commence pressure grouting and progressively withdraw the casings. The grout should be a 1:3 cement sand mix by weight. Completely fill the drill hole.
(4) After 24 hours, post-grout at slightly increased pressures above the initial pressure to enhance skin friction by improving the bond between soil and grout, and possibly to create lenses into the supporting strata. Care should be taken to avoid movement of the ground surface.
(5) In all but soft clays or peat, the surrounding soil and the frictional contact are usually sufficient to prevent buckling.
(6) Three days after completion, test all grouting.

Root (Pali Radice) piles in underpinning

This proprietary method was developed by the Italian firm Fondile of Naples in the 1950s. It consists of many double-series small diameter piles (similar to a series of angle piles) rotary drilled through existing foundations or brickwork down to a suitable bearing depth in the subsoil and bonded to the main structure with a high-strength grout compound of 700 kg of Portland cement per cubic metre of graded sand. (See Fig. 5.11.) The piles usually incorporate a single reinforcing bar for piles up to 150 mm diameter, and a reinforcing cage for piles above 150 mm diameter, and are vertically cased. The casings are extracted as the grouting proceeds and, similarly to mini post-grouted piles, compressed air pressure is used to maintain a head pressure on the grout to increase its contact with the subsoil. Piles are usually in pairs. (See Figs 5.11 and 5.12.)

A variation is the reticulated Pali Radice structure – a three-dimensional network of closely spaced piles creating a homogeneous 'stitching' of the soil and pile mass. Its main use is in 'soil nailing' land slips, and in the construction of retaining walls. (See Fig. 5.13.)

74 *Underpinning*

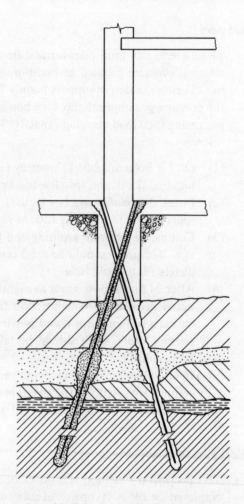

Fig. 5.11 Cross-section of Pali Radice underpinning schemes

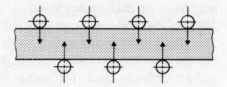

Fig. 5.12 Plan of Pali Radice scheme

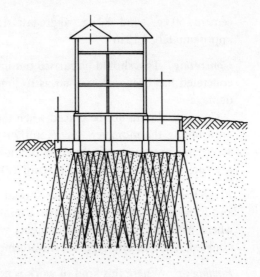

Fig. 5.13 Reticulated Pali Radice

Pali Radice underpinning is too costly for use in low rise properties, except in exceptional circumstances. The advantage of this technique is an almost immediate counteraction of movement since as soon as installation is complete the piles are unstressed and any movement in the structure causes the Pali Radice to attract load and so reduce or prevent settlement.

Specifications for underpinning work

These will vary from place to place according to practice, but the following points give an indication of the recommendations.

Concrete The specifications will cover strength, workability, type of cement and minimum cementitious content, along with references to appropriate standards. (Strengths of 30 N/mm^2 for concrete are typical.) The type and size (or grading) of aggregates and the water content or water:cement ratio should also be specified.

The minimum strength of concrete used in underpinning should be 30 N/mm^2 at 28 days for mass work and 50 N/mm^2 for concrete piles. The following mix proportions provide a useful guide for concrete with a 28-day strength of 30 N/mm^2.

Suggested mix By weight (kg/m^3), the suggested mix is OPC 380, sand 620, coarse aggregate 1190. In terms of 50 kg at mix, this would be cement 50 kg, sand 80 kg, aggregate 155 kg; and this would yield approximately $0.13\,\mathrm{m}^3$.

Concreting This should be carried out immediately after the area to be concreted has been exposed, so as to limit the risk of deterioration or damage.

Concrete should not be placed when the air temperature reaches 4°C on a falling thermometer, unless authorised by the engineer and with acceptable precautions in the form of additives and/or heating materials together with protection after placing as required.

Reinforcement Reinforcing bars should be clean and free from scale and loose rust. They should have a minimum cover of 25 mm except where used in corrosive environments when this coverage should be doubled. The size and type of reinforcement should be stated.

Pinning up Where this kind of work is needed, a normal pin-up gap of roughly 75 mm is left between the underpinning and the existing foundations. This should be completely filled with 1:3 mix of Portland cement and a sharp sand, slightly moistened and dry rammed into the gap with a rammer. Slightly moistened means that there is just sufficient water present such that the mix will bind together when squeezed in the hand. Shrink compensating additives may be used if required. Alternatively it can be flood placed using a lean concrete with small size aggregates poker vibrated into place.

Sequence of operations

The object must be to minimise disturbance both to the structure and to the occupiers of the building. Therefore, a 'method statement' outlining the sequence of all operations is important, including the safety precautions required.

Timing The time to be allowed between concreting and finishing (i.e. pinning up) should be stated (at least one day is needed). However, in practice, pinning up is often carried out as part of the concreting process using the flood concreting method.

Dimensions The dimensions of all members should be delineated (on the drawings) and the depths to which the underpinning (solid or piled) work is to be taken should be stated, with a rider making it clear that any variations in ground conditions necessitating work to greater depths will attract a variation rate. The location of position (or grid) of the piles should also be shown.

Supervision   A supervision clause is necessary. It is advisable to use the services of a competent engineer or clerk of works.

Description of the works   A detailed description of the works to be executed is usually included at the beginning of any specification and should give details of special precautions against heave if required, along with any special aspects of safety. Reference will also be made to any available information on the ground and site conditions.

Durability

The durability of concrete piles and, to a lesser extent, of steel tubes is of interest to underpinners and particularly to specifiers. The specification will certainly cover this aspect.

The use of pulverised fly ash (PFA) or fuel ash is now accepted as improving the overall performance of concrete, particularly in terms of increased resistance to chemical attack, reduced heat of hydration and reduced alkali–silica reaction.

Long-term strengths of concrete containing 25–30% selected PFA by weight of cement content exhibit a marked improvement over ordinary Portland cement concretes. This increased strength is the result of a pozzolonic reaction exhibited by calcium hydroxide (lime) depletion in the PFA concrete. This reaction causes ongoing hardening that continues over time. The increase in strength with time varies from approximately 125% of the strength of OPC concrete at 1 year to 135% at 3 years. The PFA (fuel ash) used should comply with Part 1 of BS 3892.

The depths of carbonation are similar for both types of concrete, thus equal strength mixes of each type would carbonate at similar rates.

Most studies indicate that PFA can be a useful and effective addition to concrete, generally enhancing long-term properties. In non-English speaking countries, PFA could also be referred to as *cendres volantes* or *flugenaschen*.

Infilling of piles

Steel tubes

The casings should be free from water before commencing grouting or concreting.

A normal grout would be a 1:3 OPC/sand or OPC/PFA (fly ash) mix, or a combination of all three materials. Alternatively they may be filled with flowable concrete. A minimum cement content of $300 \, kg/m^3$ is advisable. In the case of permeable sulphate strata, different mixes may be appropriate. Chapter 8 gives more details.

When a steel tube pile is filled with concrete or grout, it is the tube itself (casing) which acts as the reinforcement. The integrity of the piles depends on the preservation of the steel tubing. In a non-aggressive environment, little corrosion will occur. If, on the other hand, there is a danger of chemical attack, particularly by acid, rapid corrosion of the steel is likely. In such cases some additional steel reinforcement should be incorporated inside the casing and embedded in the concrete, preferably with a minimum of 25 mm cover in from the inner face of the steel tube. The concrete should comply with approved standards as recommended in the relevant BRE Digest guide.

Steel corrosion occurs at a semi-predictable rate – in non-aggressive environments the rate is between 1 and 5 mm per 100 years (average of 3 mm); in aggressive conditions it is 5–15 mm per 100 years (average 10 mm). These figures are very approximate and are reproduced for guidance only. However, they do indicate that a 10 mm wall thickness in aggressive conditions could last more than 100 years. Hence we can conclude that steel tubes in low rise domestic situations are generally adequate in all conditions.

The following measures can be considered to prolong the life of steel tube piles.

(1) Use a higher grade of American Petroleum Institute (API) specification steel pipe. This should increase the life expectancy of the pile by about 30%.
(2) Use cathodic protection, although this is rarely warranted in underpinning unless the building is of exceptional architectural value.
(3) Use protective coatings, bearing in mind their effectiveness will almost certainly be reduced by driving.

These measures all add to the expense and would not normally be considered in domestic underpinning unless specified.

Concrete piles

If concrete piles are used in aggressive conditions they should always be of a sulphate resisting concrete. In the case of exceptionally aggressive conditions a larger section pile than required for the loading and bearing capacity of the site can be used to provide sacrificial concrete and so prolong durability. The degree of increase will depend on experience but is unlikely to exceed 10 mm per 100 years in the case of good sulphate

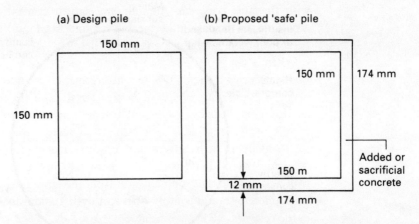

Fig. 5.14 Use of sacrificial concrete to prolong durability in aggressive ground conditions

resisting concrete. Consequently an extra 12 mm of concrete all round will be more than adequate. (See Fig. 5.14.)

With increased pile sizes it may be possible to use good OPC concrete, even though sulphate resistance is preferable, as provided in a PFA/OPC concrete, or a concrete made with sulphate resisting cement.

It is often more cost effective to use oversize piles, whether filled tube or concrete, rather than expensive coatings, additives or thicker steel.

Summary of piling solutions and usage

The causes of movement that result in a need for underpinning are shown in Fig. 5.15. Although pile solutions are appropriate in many cases, their use is rare in comparison with traditional systems. Figure 5.16 shows the proportions of underpinning work carried out using the various systems available. The reasons for the continued dominance of traditional methods are:

- perceived cost advantages;
- contractual ability; and
- ground conditions.

However, it is anticipated that piled systems will make substantial inroads into the share of traditional methods fairly rapidly in the future as the benefits of piling become more widely understood and the techniques become cheaper in line with the increased volume of work. For instance, one major advantage of piled solutions is the considerable

80 Underpinning

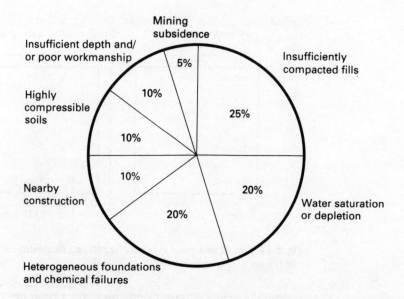

Fig. 5.15 A breakdown of the reasons for underpinning

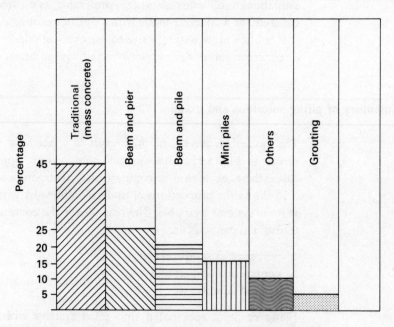

Fig. 5.16 Usage of the various underpinning systems world-wide

additional security in the case of damage caused by heave and vegetation problems.

Underpinning system and product ranges

The range of systems and products available to the underpinner are as follows.

External systems
Traditional mass concrete
Jack-down system
Cantilever pile and beam
Cantilever ring beam
Pile and knuckle beam
Pier system knuckle
Pad and beam
Extension foundations

Internal systems
Angle pile system
Double angle pile
Pile and beam
Pile and reinforced concrete
Heave systems
Pressure grouting
Slab stabilisation
Pin piled slabs and bases
Structural lifting and jacking

Standard mini-piling product range

Top driven permanently steel cased in situ concrete in piles

Size and type
65 mm diameter, circular section

90 mm diameter, Circular section

Notes
(1) 65 mm and 90 mm may be constructed using design-mix grouts
(2) Perforated tube may be used for 90 mm with a grout installed under pressure
(3) Reinforcement: single centre bar (Y12)

Bottom driven permanently steel cased in situ concrete mini piles

Size and type
105 mm diameter, circular section

150 mm diameter, circular section

220 mm diameter, circular section

Notes
(1) 105 mm mini piles may be installed uncased (reducing pile diameter to 100 mm)
(2) All mini piles may be constructed using design-mix grouts
(3) All mini piles may be pre-augered

<u>Size and type</u>
300 mm diameter,
circular section

<u>Notes</u>
(4) All mini piles may be pre-drilled
(5) 105 mm and 150 mm may be pre-drilled using Odex drilling systems
(6) All mini piles may be rock socketed
(7) Reinforcement: single centre bar (as required); cage availability in 220 mm and 300 mm
(8) 105 mm and 150 mm available with heave resistant coating

Jack-down mini piles

<u>Size and type</u>
125 mm diameter,
circular section.
Heavy wall tube

140 mm diameter,
circular section.
Heavy wall tube

<u>Notes</u>
(1) Other sizes for both circular section and universal column section available
(2) Circular sections may be concrete or grout filled as required

Augered uncased in situ concrete mini piles

<u>Size and type</u>
105 mm diameter,
circular section

165 mm diameter,
circular section

220 mm diameter,
circular section

250 mm diameter,
circular section

300 mm diameter,
circular section

<u>Notes</u>
(1) 165 mm, 250 mm, 300 mm available utilising hollow stem augering techniques
(2) All augered mini piles utilise segmental flight auger systems for reduced headroom and restricted access

Chapter 6
Pile design and piling equipment

The capacity of a pile to support a foundation is a function of the bearing capacity of the soil and the skin friction at the pile–soil interface. Thus, the calculation of the load carrying ability of a pile requires the summing of two components, namely skin friction and end bearing resistance.

A pile relying primarily on friction is, obviously, referred to as a 'friction' pile, and one receiving its support from safe, competent rock or soil strata is termed 'end bearing'. In practice, most piles derive their ultimate bearing capacity from both skin friction and end bearing resistance, the friction being mobilised first at lower applied loads.

Behavioural determinants of pile design

The bearing capacity of ground is affected by superimposed loading. In pile calculations it is usual to assume that the loads are applied relatively quickly following installation. Figure 6.1 shows the ensuing load–settlement relationship graphically. Initially, when a load is applied, the skin friction between the pile and the soil sets up a resistance to major movement as far as some point, x. Up to here the system behaves elastically (i.e. on removal of the load, the pile would return to its original level). Beyond x we reach point y, known as the yield point, at which stage the removal of the load would result in the pile adopting a set at a lower level than the original (point A). After this, any further loading is not resisted by friction, but is supported only by the bearing of the base of the pile on the substratum. Thus the bearing capacity of the soil is the key factor. When the load exerted across the sectional area of the pile base exceeds the ground's ability to resist it (point B), large-scale settlement can occur. This point is termed the 'load at failure'. The value of the failure load is a key element in pile design and should be determined as accurately as possible for the given circumstances.

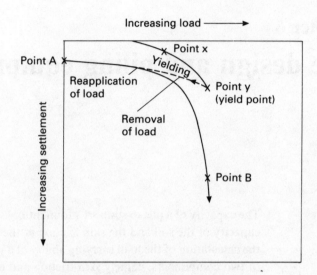

Fig. 6.1 The load–settlement relationship under progressive compression

In shallow piling and low loading – often the case in domestic work – little support is derived from skin friction, so it is usual to consider only the end bearing resistance. For short piles in cohesive soils, the ultimate end bearing resistance, Q_b, can be deduced from the formula:

$$Q_b = N_c\, C_b\, A_b$$

where N_c is the bearing capacity factor (usually taken as 9 but can be greater in very stiff clays), C_b is the undisturbed, undrained cohesion (or shear strength) of the ground at the base of the pile and A_b is the cross-sectional area of the pile base. This formula can be used when the appropriate data are available from an expert's ground investigation. If they are not, the bearing capacity of the soil must be deduced from experience and crude soil tests. In this case the calculation is done simply by dividing the load per square metre by a rough estimate of the ground bearing capacity per square metre to arrive at a cross-sectional area for the base.

For piles over, say, 2 m deep, friction support becomes significant and is, therefore, calculated and added to the end bearing support. The ultimate skin friction resistance, Q_s, is given by:

$$Q_s = a\, C_s\, A_s$$

where a is the adhesion factor, C_s is the undisturbed, undrained cohesion of the soil around the pile and A_s is the pile surface area. The adhesion factor can be obtained from tables provided in pile design manuals. It varies with strata from 0.2 to 0.8.

Finally, the permissible safe design load, Q_p, is given by dividing the base bearing resistance (and skin friction resistance, where appropriate) by a factor of safety.

$$Q_p = \frac{Q_b + Q_s}{\text{factor of safety}}$$

The factor of safety is an arbitrary factor used to compensate for the lack of a reliable method of predicting the deformation behaviour of a pile under loading. The value used for the factor of safety depends on the experience of the designers or contractors and their confidence in the empirical methods used. The figure in domestic work normally varies between 1.5 and 2.5, 2 being acceptable in most cases and a greater value usually applied if the pile is in tension.

Table 6.1 gives a rough guide to the selection of an appropriate factor of safety in relation to the quality of the data available. As the empirical test results become less reliable (i.e. unverified) a higher factor of safety is chosen. If the designers or contractors have reason to doubt the results because of previous experience of the prevailing soil conditions, caution will dictate a choice of the highest factor of safety.

Another factor to consider in the design is the length of the pile (h) in relation to its thickness (t). Here, thickness is measured along the shortest axis in the case of a rectangular pile, or as the width of one side for square piles. Large piles will receive some support against buckling from the surrounding strata, but this is ignored in the design of smaller piles (e.g. those used in domestic underpinning). For these, the designer will use only the slenderness ratio, given by h/t. The greater this ratio, the greater the tendency of the pile to distort or, perhaps, fail. Hence a round or square pile section would be preferable to one that is rectangular.

Table 6.1 The relationship between factors of safety and reliability of test results for normal loading

Method of determining capacity	Minimum factor of safety	
	In compression	In tension
Empirical prediction verified by load test	1.5	1.5
Empirical prediction verified by pile driving analyser	2.0	2.5
Empirical prediction unverified by load test	2.5	2.5

The above guidelines are subject to normal 5–10 mm settlement at working load

Example results and calculations

This section gives some representative results from in situ tests in a granular soil and example calculations to illustrate the steps in a typical design formulation process.

The shear resistance results from the standard penetration and static cone resistance give a measure of the ultimate bearing resistance. Some typical results are shown below.

Standard penetration resistance (blows/300 mm)	Angle of shearing resistance (degrees)
10 ⎫	30.0
20 ⎬ in medium dense granular soil	33.0
30 ⎭	36.0
40 ⎫	38.5
50 ⎬ in dense granular soil	41.0
60 ⎭	43.0

Static cone resistance method (kg/cm^2)	Angle of shearing resistance (degrees)
50	35.0
75	36.0
100	37.5
150	39.0
200	40.0

It will always be difficult for non-specialist contractors to deduce the end bearing resistance and skin friction from the equations shown earlier because of the difficulty in obtaining an accurate value for C_b and C_s. These will only be available after an expert ground investigation so cruder practical calculations are often the only option. Some guidance on these can be found in British Standards. For instance, BS 8004 (Foundations) defines the ultimate bearing capacity of a concrete pile as the load at which the resistance of the soil becomes fully mobilised. This is generally taken to be the load which causes the pile to settle to a depth of up to 10% of its width or diameter and equal to 0.25 the compressive strength of the concrete. For precast concrete piles, BS 8004 also gives standards for the usage of longitudinal steel reinforcement to cater for stresses imposed by lifting, handling and tensile loading. The requirements are shown in Table 6.2. Guidance on the maximum permissible loads which can be imposed on various soils is also given in this Standard.

Table 6.2 BS 8004 requirements for longitudinal steel reinforcement of precast concrete piles

Volume of steel at head and toe	Volume of steel in pile shaft	Steel cover	Other requirements
0.6% gross volume over a distance of 3 × pile width from each end	0.2% gross volume spaced at not more than 0.5 × pile width	20 mm fast grade 50 concrete 25 mm to grade 40 30 mm for grade 30	Lapping of short bars with main reinforcement to be arranged to avoid sudden discontinuity

Example calculations

A precast concrete pile is required to carry a compressive working load of 300 kN and 200 kN in uplift. It is to be driven into soft clay which has an undrained shear strength of 100 kN/m² and an adhesion factor of 0.63. A safety factor of 1.5 has been specified and the proposed design is a 250 mm square pile driven to a depth of 6.5 m. To confirm that the design is satisfactory, we follow the steps below.

(1) End bearing $(Q_b) = 9\,C_b\,A_b = 9 \times 100 \times 0.25^2 = 56.25$ kN
(2) Required skin friction in compression
$(Q_s) = 450 - 56.25 = 393.75$ kN

For a design pile depth of 6.5 m

(3) $Q_s = a\,C_s\,A_s = 0.63 \times 100 \times (4 \times 0.25 \times 6.5) = 409.5$ kN
(4) Total pile resistance $= Q_b + Q_s = 56.25 + 409.5 = 465.75$ kN
(5) Factor of safety in compression $= 465.75 \div 300 = 1.55$
(6) Factor of safety on uplift (Q_s only) $= 409.5 \div 200 = 2.05$

A pile depth of 6.5 m therefore satisfies the specified factor of safety of 1.5 in both compression and uplift.

Heave calculation In a heave situation in clay the calculations are the same as foregoing but additionally provision is required for a sleeve, slip coat or mobile fill around part of the pile (see Chapter 2).

Plant and equipment

Underpinning generally involves working in areas where access is restricted and headroom is limited. This is particularly the case when operating in domestic properties or close to adjacent structures. These physical constraints are a major factor in the design of an underpinning solution using piles. Such piled systems could involve the installation of driven, bored or augered piles, from 60 mm to 300 mm in diameter, under conditions where the minimum headroom is around 1800 mm and access is restricted to around 750 mm. To overcome these problems, conventional piling plant has been adapted to the working conditions and purpose-built equipment has been produced. The following paragraphs outline the machinery used by specialist underpinning contractors.

Grundomats

These compressed-air tools are used for bottom driving small diameter (e.g. standard 60, 90 and 105 mm) steel tube mini piles. The equipment illustrated can be manhandled and used in confined spaces. It is proprietary plant, made and marketed by the Grundomat Company, and was originally developed for installing underground public utility services, mainly by horizontal boring. It has now been adapted for vertically installing mini piles – see Fig. 6.2.

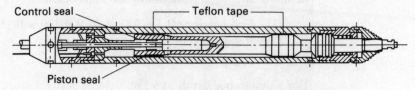

(a) Section through Grundomat compressed-air tool

(b) Grundomat in operation

Fig. 6.2 The Grundomat compressed-air tool

Drop hammer rig

This method is used for installing steel tubular piles of diameters between 100 and 200 mm by means of bottom driving. In essence, a heavy weight is suspended from an A-frame and winched up to drive down on to a dry concrete plug at the foot of the pile. The equipment can be mounted on tracks, as shown in Fig. 6.3.

Fig. 6.3 Drop hammer rig on tracks

Top drive hammers

Compressed-air driven, these are modified jackhammers and are used for driving steel tube up to 100 mm diameter. Figure 6.4 shows a top drive hammer in operation.

90 *Underpinning*

Fig. 6.4 Top driving a mini pile with a compressed-air vibratory hammer

Drill/auger rigs

These rigs are used for removing the ground either by drilling or augering prior to forming a cast in situ or continuous flight augered (CFA) pile. They are also used for pre-augering or drilling when installing driven piles. (See Fig. 6.5.)

This sort of equipment can be divided into two general categories:

(1) *Standard rigs:* These are manhandleable and can be moved around the site by means of a set of wheels. They are secured to 'ground' when in operating mode and are ideal for installing small diameter piles (up to 250 mm) in situations where access is restricted. (See Fig. 6.6.)
(2) *Tracked rigs:* Ostensibly the same equipment but mounted on tracks for greater mobility and enhanced stability in operating mode. They are suitable for piles up to 450 mm in diameter. (See Fig. 6.7.)

There are, of course, other proprietary drilling rigs besides the purpose-designed Roger Bullivant equipment pictured here. For example, the

Pile design and piling equipment 91

Fig. 6.5 Drill/auger rig

Compair Holman M-Trak is a compact hydraulic crawler drill that is capable of drilling holes 60–150 mm in diameter down to 35 m and 150–300 mm down to 15 m in depth. The track width is adjustable between 700 and 900 mm. It is powered by a shed mounted hydraulic power pack

Fig. 6.6 Standard drilling rig being moved around the site on wheels

92 Underpinning

Fig. 6.7 Track mounted drilling rig

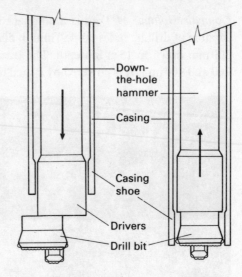

Bit open drilling
The eccentric bit rotates outwards drilling a larger diameter hole than the following casing

Bit closed pulling
The eccentric bit rotates inwards and the drill string is ready for pulling

Fig. 6.8 Example of a down-the-hole hammer

Pile design and piling equipment 93

set remote from the drill. An alternative from the same company is the *Uni-Trak* machine.

The wide range of hole sizes that can be drilled is achieved by using augers, 'down-the-hole' hammers and casing systems as appropriate. The principles of down-the-hole hammers are illustrated in Fig. 6.8.

Jack-down rigs

Using this equipment, piles are installed by hydraulically pushing short sections into the ground. A jacking frame is secured to an adequate base and hydraulic rams impose the load on to the piles. This system is ideal for work in areas of restricted access where vibration could be a problem. Figure 6.9 shows such a rig in operation.

Fig. 6.9 Installation of jack-down piles

Chapter 7
Grouts and grouting

While cementitious grouts are used as an alternative to concrete for infilling steel and tube piles, grouting (both cementitious and chemical) has other uses, namely strengthening ground to avoid the use of piles altogether, or improving the ground to upgrade an underpinning system. In effect, grout can stabilise ground, not only by reducing soil permeability, but also by increasing its strength.

Ground improvement by grouting

The improvement of ground by enhancing impermeability and/or strengthening, and so improving bearing capacities, can be achieved in several ways by grouting.

(1) The process of *permeation grouting* uses a grout to fill most of the voids in the ground without significantly changing the original volume of the soil. The grout must be fluid enough to penetrate pores and fissures under controlled and low injection pressures.

(2) *Compaction grouting* uses higher pumping pressures and a stiff grout to create a growing bulb at the point of exit, thus compacting the surrounding strata by compression and rendering the soil less permeable, which in turn increases its bearing capacity. This technique is useful in correcting differential settlement and uplifting settled solid floors.

(3) During *Clacquage* or *hydrofracture grouting*, high pressures fracture the ground and the grout coats and compresses the individual fragments. In addition, lenses of grout are pushed out into weak areas, without encapsulating soil fragments in matrix material. The very high pressures under which the grout is injected mean care is needed to avoid disruption of the surface. A variation of this technique is squeeze grouting. These types of grouting require expertise and should be carried out by specialists.

(4) *Jet grouting* is a process in which the soil is mixed in situ with a stabilising mixture under high pressure.

These grouting techniques offer an alternative method of counteracting damage caused by ground movement and can be used to upgrade failed underpinning work by stabilising the underlying strata. The appropriate technique and grouting material will be dictated by the circumstances and, frequently, by the budget available for the work. The grouts available fall mainly into two categories – cementitious and chemical. These are described in more detail in the following sections. The fundamental difference is that chemical grouts can permeate through soils with much finer grain sizes. However, they are much more expensive and so are often used in combination with cementitious grouts to reduce the costs.

In underpinning work, cementitious grouts usually consist of Portland cement, sand, bentonite, pulverised fly ash (PFA)/fuel ash and other additives, in various combinations.

Selecting a grout system

In selecting a grout system the following factors should be considered.

- The use: is the grout to be used for pile infilling or to strengthen ground?
- The availability of the grout constituents along with the ability to mix and place satisfactorily.
- The extent of available information on the permeability and porosity of the strata to be grouted. This will also involve consideration of the diameter, depth and extent of reinforcement to be grouted.
- The strength and stabilisation durability of the grout in situ.
- The chemical composition and quantity of the ground water.
- The characteristics of the soil itself as regards sulphates and other chemically aggressive constituents.
- The degree of site supervision available.
- The overall cost of materials, mixing and placement.

The permeability of the ground is the most important factor in selecting the optimum grout, with cost affecting the final decision. Cementitious grouts are preferable for grouting tube piles and for ground where pores are of a diameter above 0.1 mm (i.e. medium/coarse sands). Below this size, down to 3 microns, chemical grouting is necessary. Below 3 microns, chemical grouting also becomes too costly. In terms of cost effectiveness and flexibility, jet grouting offers the most attractive option in the majority of cases. It is described in more detail in the next section.

Jet grouting

Jet grouting techniques make it possible to treat any type of soil, from coarse to very fine grained (i.e. gravel to clays), with cementitious grouts. This ability removes the need for expensive chemical grouting and explains why jet grouting dominates the field.

Jet grouting involves the application of cementitious grouts through high-speed jets at high nozzle pressures of between 20 and 70 MPa. The ground is mixed with the selected grout in situ, although in other circumstances the soil can be partially removed by air/water jetting and replaced continuously after mixing with grout. The method is illustrated in Fig. 7.1.

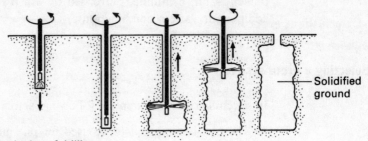

1. Beginning of drilling
2. End of drilling
3. Beginning of lateral jetting
4. Revolving and drawing up
5. Completion

Fig. 7.1 The jet grouting method

The injection equipment is fitted at its base with two pairs of diametrically opposed nozzles. Once in place after drilling, the equipment starts grout jetting by revolving as it retreats up the drill hole. The grout, under pressure, fractures and enters the surrounding ground and stabilises it as it sets and hardens, forming a wide column. The column width depends on the type of ground, the grout, its discharge pressure, the speed of nozzle rotation, the number and size of the nozzles and the updraw or lift rate.

The system was developed by Radio SA of Milan and is known as the *Radinject* system. An example of how this jet treatment can be used to improve the soil bearing capacity of shallow foundations is shown in Fig. 7.2.

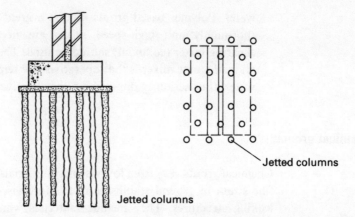

Fig. 7.2 Layout of Radinject treatment to improve soil bearing capacity of shallow foundation

Cementitious grout blends

Cementitious grouts can be subdivided into the following categories. Although a lot of grouting is carried out using neat cement, the types described below offer certain advantages,

- *General purpose grouts* are based on blends of Portland cement to BS 12 and PFA to BS 3892 Part 2A. These grouts give strengths proportionate to the amount of PFA present, ranging between 1 N/mm^2 and 28 N/mm^2 at 28 days, and 1.5 N/mm^2 and 40 N/mm^2 at 90 days. The yields of these grouts are greater than for OPC and range from 0.79 to 0.97 cubic metres per tonne. They are placeable in water, with total solids ratios of 0.35 to 0.45 depending on the mixer used. Produced in a factory, they are available in bulk tankers or in 25 kg sealed paper sacks.
- *Clay grouts* are combinations of various clays with an OPC/PFA base (1:2). The clays used are kaolinite, calcium bentonite and sodium montmorillonite They form a class of shrink compensated grouts useful in grouting sleeved pipes. They have a degree of water proofing and are sulphate resistant. The 28 day strengths are around 15 N/mm^2 with a 90 day strength of about 27 N/mm^2. Yields are of the order of 0.82 m^3/tonne.
- *Polymeric grouts* are those which contain a long chain, high molecular weight polymer and are designed to reduce wash-out in very wet conditions. Strength averages 19 N/mm^2 and yield is 0.82 m^3/tonne. They are used where water is a problem and, in some cases under-

water. Polymer based grouts require a great deal of energy to mix thoroughly and high-speed mixes are necessary. A high-speed shearing mixer (colloidal) should be used. The great danger in using an inadequate mixer is that operatives are tempted to overdose with water, thus seriously diminishing the effectiveness of the polymer.

Chemical grouts

Chemical grouts vary from low viscosity materials, for the penetration of fine strata in ground stabilisation, to more viscous grades for sealing leaking structures. They include materials which retain a degree of flexibility when set for use where movement may occur.

The selection of the correct grout, and the proportioning of the constituent materials, calls for wide experience in the properties of these materials. Failure to appreciate the contribution of each constituent to the end product can result not only in structural weakness but also in expensive wastage of both labour and materials.

Chemical grouts are normally reactive liquids and either one or two shot grouts. Due to their nature, they can be injected into much smaller fissures than cementitious grouts and can, therefore, stabilise a wider range of soils, including fine grained sands and silts.

An interesting chemical grout is the TACSS system. Based on a polyurethane type material, this liquid grout has several unique features. The basic resin is a low viscosity liquid which is easily mixed with a small percentage of catalyst. The amount of catalyst can be varied to suit to the exact nature of the application.

The type of mixer required depends on the type of grout being used and can vary from quite small chemical mixers to large double-drum Putzmeisters or high-speed colloidal mixers.

Flowing concrete

While not strictly grouting, flowing concrete is often included under the general heading 'grouting'. Accordingly, a brief reference is included here.

The labour costs of placing concrete led to interest in flowing concrete, that is, concrete with increased workability (flow), beyond the limits of normal mix designs and having a slump greater than 200 mm (or the DIN 1045 spread test of between 55 and 62 cm), while retaining the characteristic compressive strength of conventional concretes.

The use of super plasticisers has opened up new opportunities. Basically, these products fall into three groups, namely lignosluphonates, melamine formaldehyde sulphonates and napthalene sulphonate.

Design

The starting point for designing flowable concrete is to design for a pumpable mix, initially using a fine sand and a graded coarse aggregate of 20 mm or less. Similar densities and strengths are obtained with or without superplasticisers. The following mixes are typical for use in underpinning:

- 300 kg OPC, 35% medium-fine sharp sand; 65% 20–25 mm crushed stone.
  0.6 water:cement ratio, 0.6% of cement content as superplasticiers (approximately 1:4.5 OPC to aggregate).
  Strength at 28 days, 30 N/mm^2; density 2450 mg/m^3.
- 300 kg OPC, 40% medium-fine sharp sand, 60% graded 20 mm gravel
  0.6 water:cement ratio, 0.7% superplasticiser (approximately 1:4.5 OPC to aggregate.
  Strength at 28 days 30 N/mm^2; density 2425 kg/m^3.

Plasticisers may also be used with sand/cement grouts or mortars.

Advantages

The use of flowable concrete has gained favour with contractors as a grouting option for the following reasons.

- It can be placed quickly and requires fewer operatives.
- It reduces the risk of damage by vibration.
- It achieves complete compaction.
- It can be pumped quickly over long distances to give easy access to difficult areas.
- It can be placed easily around congested steel reinforcement.

Chapter 8
Safety and legal aspects of underpinning

Safety

Much of the UK safety legislation applicable to the construction industry also applies to underpinning, and has been in operation since the 1960s. In 1974 the Health and Safety at Work Act came into force and created the Health and Safety Commission whose duty it is to consult with relevant bodies and review, amend or produce new regulations as necessary. A wide range of regulatory legislation, referred to elsewhere in this chapter, affects underpinning work but three sets of regulations, dated between 1961 and 1966, are particularly significant. They are:

- The Construction (Health and Welfare) Regulations 1966
- The Construction (Working Places) Regulations 1966
- The Construction (General Provisions) Regulations 1961.

Under Section 50 (3) of the Health and Safety at Work Act 1974 the Commission issued a consultative document in 1995 proposing the implementation of the EC Directive on Temporary or Mobile Sites and the consolidation of the three foregoing regulations into a single unified code to be known as the Construction (Health, Safety and Welfare) Regulations likely to become law in 1996. These Regulations will clarify construction legislation into two main complementary sets, namely:

- The Construction (Design and Management) (CDM) Regulations 1994 and
- The Construction (Health, Safety and Welfare) Regulations (not yet enacted).

In essence, this legislation imposes a duty of care to ensure, first, the satisfactory and proper design and management of a construction project, and second, the health, safety and welfare of work people, staff and any members of the public visiting, or in the vicinity of, the site.

Several authorities may become involved in the event of damage or accident occurring, such as the factory inspectorate, building control authorities, environmental officials and public utilities, who have powers of inspection and prosecution in the case of serious infringements or the ability to close down the work until infringements are remedied. Examples include:

- Reportable accidents
- Dangerous working conditions and practices
- Unsafe buildings
- Damaged utilities, particularly gas, water and electricity
- Works involving listed buildings, ancient monuments or premises in conservation areas
- Work affecting trees covered by preservation orders
- Work affecting highways and sewers.

The CDM Regulations 1994 have placed a duty on designers and contractors to assess any hazards and risks which may be encountered during the progress of the works, including such items as the need for shoring, timbering, pre-repairs and the depth of working.

The CDM Regulations

The CDM Regulations apply to the construction industry, including underpinning firms, who employ five or more people. They apply to clients, designers, what the Regulations term 'principal' contractors and subcontractors.

Of particular importance to underpinners is the need to carry out a risk assessment. This applies to clients, designers and principal contractors, who have responsibility for ensuring subcontractors also comply. There is an exception for domestic householder clients, but this does not extend to the designer or underpinner working on domestic property. A risk assessment must identify all foreseeable hazards to health and safety of employees and all other persons connected with or likely to be affected by the proposed works.

The Regulations require the appointment of a 'planning supervisor', being a competent person who must prepare for each project a 'health and safety plan', similar to earlier method statements but wider in scope, to monitor all health and safety aspects and risks of a design and to require the provision of adequate resources of people, material and finance in order to satisfactorily complete the project. The compilation and keeping of a 'health and safety file', as discussed later, is mandatory.

Designers must design to avoid, reduce and control identifiable risks so far as is reasonably practicable, and to enable a competent contractor to execute work safely, satisfactorily and without injury to the health of the workforce.

The principal contractor is required to adopt fully the health and safety plan and to co-ordinate the work of subcontractors (such as underpinners) to ensure compliance with the plan and the law.

The health and safety file must contain information of help to people concerned not only during but also after completion of the work, such as final drawings showing the work as actually constructed, the design calculations, any variations, general details of materials used, subsequent maintenance and operational instructions, and the location of public utility installations affecting the works. Apart from being a mandatory requirement, such a record is likely to be of use in the event of claims for defects liability, litigation, recalls or further work in the vicinity. The health and safety plan required by the Regulations should contain:

(1) Client details, location, nature and description of the works required and the timescale for their completion.
(2) Brief details of the environment, including location of public utility services, sewers, access requirements (particularly for fire and ambulance services), adjacent structures and any precautions required by these, any special health and safety problems, ground conditions and other identifiable risks.
(3) Notes on the risk assessment and who carried it out, along with the design principles adopted to cover the precautions to be taken during the progress of the works.
(4) In the case of materials, details of any health hazards possible from the use of specified materials and their handling and storage.
(5) Site and working standing orders being specific rules of working, permits to work and any emergency procedures.

The object of these Regulations is to develop and encourage a health and safety culture in all construction firms so as to ensure projects are carried out safely and effectively and without injury to the health of the people employed and the general public.

An approved Code of Practice, 'Managing Construction for Health and Safety', has been issued by the Health and Safety Commission to provide guidance on complying with the CDM Regulations. A similar code exists for the Management of Health and Safety at Work. The following guides and codes applicable to the industry are available from HMSO.

- Managing Construction for Health and Safety (ref CDM Regs 1994)
- Designing for Health and Safety in Construction (ref Health and Safety at Work Regs 1992)
- Management of Health and Safety at Work (ref MHSW Regs 1992)
- Workplace Health, Safety and Welfare (ref WHSW Regs 1992)
- Personal Protective Equipment at Work (ref PPEW Regs 1992)
- Manual Handling (MHO Regs 1992)
- Work Equipment (WE Regs 1992)
- Display Screen Equipment Work (H & S (DSE) Regs 1992)

These should be in the possession of designers, contractors and anyone concerned with the subject.

Difficulties associated with underpinning

By its very nature underpinning is often carried out on unstable structures with the attendant danger that the underpinning work may aggravate the condition. It is imperative therefore that measures are adopted which ensure the stability of the building and these are dictated by:

- The choice of technique used
- A limit on any excavation open at any one time
- A reduction in loading on the foundation by internal propping and/or shoring the property
- Regular, competent supervision
- Adequate working space.

Work in confined spaces

This usually applies to mass concrete work or underpinning carried out in basements. The main risks are:

- Unsafe excavations
- Toxic gases and/or deficiency of oxygen.

All work carried out in confined spaces should provide for regular air checks to detect any harmful change in conditions. This is important when using mechanically driven plant or welding apparatus.

Working with or near electricity

Prior to the commencement of underpinning working and where there is a probability of encountering electrical lines or apparatus, notification should be given to the electrical supply authority along with a request for information on the location and depth of any buried cables, or precautions needed regarding live overhead lines.

The Electricity at Work Regulations 1989 apply and it should be noted that electricity for powered tools must be 110 V a.c. or less. In confined spaces it is advisable to use specially designed, battery driven equipment.

Health hazards

The ones associated with underpinning are noise, micro-organisms, deleterious materials and incorrect lifting of loads. The UK Noise at Work Regulations 1989 set the limits acceptable for drilling and piling. Where the permitted levels are likely to be exceeded, muffling, attenuating exhaust air of equipment by use of thick wall plastic tubing, or reducing operating air pressures, are all methods used to reduce noise. Ear protection should also be supplied.

Deleterious materials can be protected against by the use of gloves, protective clothing and, if necessary, masks and by providing ready to use washing facilities. Welding and dust fumes can be reduced by proper ventilation or in extreme cases by providing masks. Protective headgear is an essential requirement.

Vibration damage

When other construction work is occurring adjacent to or near property damaged by subsidence and being repaired by an underpinning system, or heavy traffic or vibrating machinery is operating, vibration can occur and precautions to minimise or control such effects need to be taken by way of timbering. It should also be borne in mind that underpinning works themselves such as piling can be a source of vibration and be the subject of a local authority control order or even a court order.

Main legislation affecting underpinning work

The following are the main items of legislation which concern underpinning work:

- Health and Safety at Work Act 1974
- Building Regulations 1991
- Building Regulations (Amendment) Regulations 1994

- Health and Safety (Enforcing Authority) Regulations 1989
- Construction (Design and Management) Regulations 1994
- Construction (General Provisions) Regulations 1961 (under review)
- Construction (Working Places) Regulations 1966 (under review)
- Construction (Health and Welfare) Regulations 1966 (under review)
- Construction (Head Protection) Regulations 1989
- Protection of Eyes Regulations 1974
- Health and Safety (First Aid) Regulations 1981
- Control of Substances Hazardous to Health Legislation (COSHH) 1988
- Electricity at Work Regulations 1989
- Noise at Work Regulations 1989
- Handling Operations Regulations 1989
- Construction (Lifting Operations) Regulations 1966
- Workplace (Health, Safety and Welfare) Regulations 1942
- Latent Damage Act 1986
- Control of Pollution Act 1974
- Local authority building bylaws.

It is apparent that careful planning and a knowledge of current regulations are needed at the outset, not only to define the method of working, the design and the materials, but also to ensure compliance with the mass of regulatory matter.

Temporary supports for premises being underpinned

A major concern in underpinning work is the possibility of collapse. Where underpinning is to be executed to limit or arrest movement due to settlement, as in traditional or mass concrete underpinning, pre-surveys often indicate the need for shoring, strutting, propping and trench timbering. The techniques and recommendations are outlined below.

Strutting and shoring

Strutting is a simple operation and consists of interposing timber or steel struts and braces to windows and door openings to prevent further misalignment as work progresses.

Shoring involves constructing timber or steel supports to walls. There are three main techniques in use:

(1) *Needle and dead shoring* This type of shoring is used for the supporting walls during underpinning and involves the installation

106 Underpinning

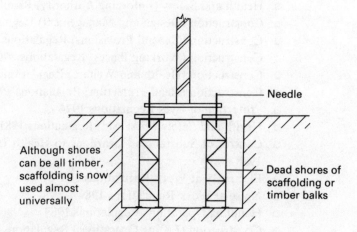

Fig. 8.1 Needle and dead shore

of vertical members either side of an excavation, supporting needles spanning across them.

The upright or dead shores may be timber bulks, scaffolding or steel stanchions – Fig. 8.1.

(2) *Raking shoring* These are used to support the exterior walls of a building likely to be affected structurally by any underpinning operations.

They comprise tubular steel or timber members set, in accordance with good practice, at angles not exceeding 70° to the horizontal and so inclined to thrust against the wall being supported from a sole plate, spreading the reaction over an area of solid ground – see Fig. 8.2.

The top of each shore pushes against needles let into the wall immediately below each floor position and fixed against upward movement by a cleat fixed to the wall and to the needle. The thrust at ground level is taken by a sole plate of a size capable of distributing the anticipated load over the bearing capacity of the ground and clear of the area needed for any excavation (at least equivalent to 50% of the depth of the excavation), unless adequate trench supports designed to resist the loading caused by the shoring is provided.

(3) *Flying shoring* Where possible support is available from adjacent buildings (i.e. within 10 m) and where raking shores would cause obstructions, flying shores are an alternative. They consist of vertical timbers secured to both opposing walls and needled and cle-

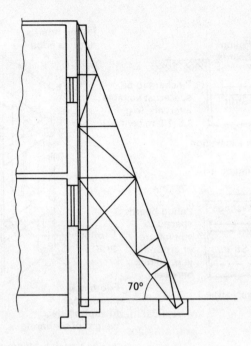

Fig. 8.2 Raking shore

ated as for raking shores. They should be set across floor levels wherever possible, or backed up by internal strutting, with struts or scaffolding spanning between the vertical timbers – see Fig. 8.3.

Flying shores are used to prevent bulging or outward movement of walls.

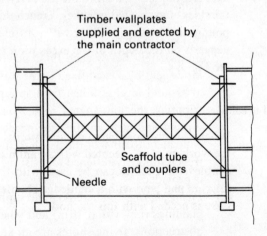

Fig. 8.3 Flying shore

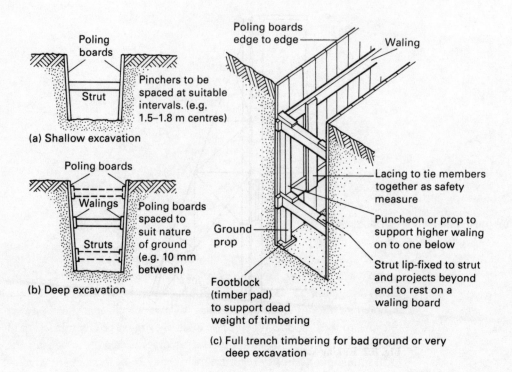

Fig. 8.4 Trench timbering

Trench timbering

This technique is similar to the use of flying shores but is used to support the sides of excavated trenches. Trench timbering makes use of strutting, polings and walings, with or without trench sheeting. The configuration depends on the depth of the excavation and the nature of the ground – see Fig. 8.4.

Cantilevered needles

Where it is not possible to obtain internal access to a building (e.g. when the work must be executed with a minimum of disruption to occupants), cantilevered needles can be used, with the needles being support on a fulcrum pad ;and with a kentledge counterbalance weight – see Fig. 8.5. Care is needed with this method.

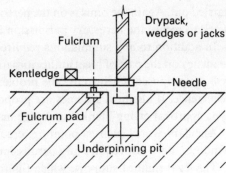

Fig. 8.5 Cantilevered needle

Approval of underpinning designs

In England and Wales building controls are governed by the Building Regulations 1991 as amended by the Building Regulations (Amendment) Regulations 1994. In addition, some local authorities have adopted or modified model building bylaws which give them powers relevant to local needs. When these exist, they must be complied with in addition to national regulations, so it becomes necessary to consult the local municipal authority regarding what may be in force in any area of intended operations before embarking on an underpinning scheme.

Building Regulations in the inner London area require special attention owing to their complexity and the differences compared with requirements elsewhere. Similarly, different rules apply to Scotland and Northern Ireland.

Building Regulations and bylaws cover foundation designs, which includes underpinning, and refer extensively to British Standard Specifications and Codes of Practice. A useful definition which illustrates the scope of this coverage is:

> A foundation or an alteration to an existing foundation must safely sustain and transmit to the sub strata, the combined total loading on the structure in such a manoeuvre as not to cause damage to any part thereof, or to any adjoining or nearby properties, and should be taken down to such a depth and be designed in such a manner and be so constructed as to safeguard the building against further movement damage by heave, shrinkage or freezing of the soil, and be capable also of resisting chemical attack from the surrounding soil or present in ground water.

Many local authorities require plans and calculations for underpinning schemes, while others only require formal notification that work is to be carried out. Again the onus is on the persons authorising or executing the work to secure the necessary permission if required.

In addition to any submissions required, there is today an increasing tendency on the part of local authorities to seek satisfaction that not only will the proposed scheme cure the problem, but also that it will prevent recurrence. If identical schemes are proposed, as for say an estate, then type approval giving blanket coverage may be possible.

Ancillary work, such as pointing, crack repair, eaving of doors and windows, does not require any permissions.

The legal right of support, which figures in all underpinning work, is rightly the province of the legal profession and it is advisable to consult a solicitor (see later in this chapter for a brief summary of the legal position) when this type of problem occurs. An alternative is to consult a specialised civil engineer: advice on persons specialising in underpinning can be obtained from the *Association of Consulting Engineers Handbook*.

Some specialist underpinning contractors can draw on previous experience and provide sound advice but without the authority of legal training. The *Geotechnical Directory of the UK* (British Geotechnical Society/Institute of Civil Engineers, 1989–90) can also provide guidance on specialists.

Guarantees

In the event of underpinning work failing, a client may have grounds for suing either for breach of contract or for negligence. However, a client may prefer to rely on the protection of a guarantee, guarantees or warranties being increasingly available to the purchaser of products or services in almost any market sector in industrialised countries.

The guarantee or warranty offered in connection with underpinning work may depend on which party in the transaction is offering it, the scope of the work covered, and the contractual position between the parties. In a typical underpinning contract the following parties are involved.

- The client
- The client's insurance company
- The insurer's appointed loss adjusters
- A consulting engineer approved by the loss adjuster
- The contractor appointed by the consulting engineer and possibly a subcontractor appointed by the contractor
- Any supervisor or clerk of works appointed.

In some circumstances a guarantee may be issued by the contractor direct to the client covering the works undertaken.

Typically the guarantee will stipulate the time period to be covered, not exceeding 15 years and probably normally 10 years, and include a very general wording of the extent of the contractor's obligations, which could be open to varying levels of interpretation. Beware, for example, the contractor who guarantees to repair or reinstate the works if he 'is found to be liable'. The contractor, in particular circumstances, might argue that the design, if carried out by another party to the works, is faulty, and that may in turn involve the consulting engineer. This in turn could depend on whether the problem was correctly appraised and whether the solution was designed appropriately.

The real value of a guarantee depends on how specifically it is worded in terms of addressing liability for the effectiveness and satisfactory nature of the work done. It also depends on who is giving the guarantee.

Undoubtedly the best form is that underwritten by an independent insurance company and one which is transferable to a subsequent owner in the event of the property changing hands, provided always that such a transfer is within the time limit of the policy. The policy should also state that it remains valid and in force should the contractor become insolvent or cease trading. Otherwise the policy, like a non-insured guarantee, would be worthless.

It always pays to use only reputable, long established specialist underpinners who will usually offer to provide an appropriately insured guarantee. Consultant engineers and contractors also carry public liability policies and professional indemnity cover. So, wherever possible look for an independent insurance backed guarantee of sufficient duration to meet required needs. However, some physical work is either uninsurable or extremely expensive for example, bad workmanship, which often is covered by an exclusion clause even in otherwise good policies. This can be overcome by a contractor guarantee, backed by deposited monies. Again, if in doubt seek legal advice.

Suitable insured guarantees are costly and not always obtainable. In the UK there is an association of specialist underpinning contractors who provide a way of obtaining additional security by ensuring the capabilities of their members through a code of practice. It is also good practice to use contractors who are assessed for quality assurance under the BS/ISO/EN system.

Right of support for adjacent landowners

Guarantees and public liability policies may provide cover for damage to adjacent land. An adjacent landowner will have the right to require that any use of adjacent land does not withdraw his right of support from the land.

Normally each portion of land gives 'natural support to the other portions in both the vertical and horizontal planes. If underpinning disturbs this right, causing damage, litigation could follow. This does not strictly apply as a natural law to structures erected on the land which have acquired rights by the passage of time. Such rights can only be tested by the courts and this is the province of a lawyer.

Contracts

The essence of any contract is that two parties agree a bargain whereby one party undertakes to provide some consideration or payment in exchange for a product or service offered by the other party. In the case of underpinning work, the contract should be in writing. The party who pays is called the client, or employer or purchaser, and the other party the contractor. Obviously, there has to be a clear understanding between the parties to a contract as to what is required to be performed in return for the agreed price.

The client often appoints a representative to act for him, such as a consulting engineer who may design the scheme or specify the materials and workmanship for the contractor to price. Alternatively, it may be that the contractor is required to design and submit the scheme for the approval of the consultant. Whatever method is used, the contract should state in clear, concise and unambiguous detail the client's requirements and the sum to be paid. The contract documents are likely also to comprise plans, sections, specification, method statement and written details of how the cost is to be ascertained. It is usual to use a standard form of contract including either one of the Joint Contracts Tribunal standard forms of building contract or the Institution of Civil Engineers' Conditions or Minor Works Agreement:

- JCT 80 Private edition with quantities
- JCT 80 Private edition without quantities
- JCT 80 Local authority edition with quantities
- JCT 80 Local authority edition without quantities
  Note: The above can all have a contractor design option.
- JCT 81 with contractor's design
- JCT 84 Intermediate Form
- JCT 80 Minor Works Agreement

- ICE Conditions 6th edition
- ICE Minor Works Conditions.

The more popular forms used in underpinning are the JCT 80 Minor Works Agreement or ICE Minor Works Conditions. On very small, domestic jobs when building contractors are employed the work may sometimes be carried out using a written quotation (not an estimate), either with a fixed price or a clearly defined way of determining a costing variation. These types of contract would not be acceptable to an insurance company. Before setting the final account, some form of certification of satisfactory completion and conformance with plans and specification should be required. This should preferably be from a qualified professional person.

Finally, any contract should contain clauses covering:

- Clear and concise details of the client's requirements
- Verification before final settlement of satisfaction and conformation with the contract documents
- Full recording of all transactions between the parties during the progress of the contract
- A defects liability period during which the contractor will remedy any defects.

Civil liability for foundation failure

In the UK the position as regards liability for negligent design or workmanship is briefly as follows.

Negligence: The definition

Negligence, which is a tort, can be defined as a breach of any obligation at common law, or of statutory duty, and/or arising from the express or implied terms of a contract, to take and exercise skill and care in the performance of a contract or legal duty.

There have been considerable changes in the law regarding negligence in recent years but before there can be liability there must be:

- A duty to take care
- A breach of that duty
- Damage suffered by the plaintiff in the form of personal injury or physical damage to his property or, in the case of negligent statements, financial loss alone as a result of the negligent statement

- 'Proximity' of plaintiff and defendant
- In addition to 'foreseeability' of damages, it must be 'fair, just and reasonable' for the court to make an award of damages.

Designers, engineers, architects, surveyors, contractors, developers and local authorities can all be sued for negligence, either individually or in combinations.

The liability of the contractor

The contractor's liability is to provide works which use proper materials, are carried out in a good workmanlike manner and are fit for the purpose intended, which in underpinning means the safeguarding of the premises from future foundation damage and leaving the premises fit for habitation. This is a common law duty and also a statutory duty under the Defective Premises Act 1972.

Professional people (designers, engineers, architects) have a duty to their client. This duty is to use all reasonable care and skill in the course of their work and where an engineer is engaged to provide design services for a contractor, a fitness for purpose may, in the absence of express agreement, be imposed. Conforming to the Building Regulations and following the best technical practice offers the best defence.

Limitation

For an action based on negligence, a writ must be issued within 6 years of the cause of the action having arisen. This means within 6 years of the date at which the damage occurred.

The Latent Damage Act 1986 has increased to 15 years the time period within which actions may be brought for damages for negligence not involving personal injury. Time starts to run from the time when the work, including remedial work, is done.

Remember always that the onus is on the defendant to prove the defence and that the sympathy of the court is usually with the plaintiff. Thus, always use competent contractors, adhere to codes of practice, BRE Digests and good current practices, and always seek professional legal advice when a claim is likely to be made.

Nuisance

Nuisance is a tort, quite distinct from negligence, and consists of an inconvenience or 'nuisance' inflicted by the defendant, or the plaintiff, which materially affects the plaintiff's enjoyment of his property.

If underpinning work does damage to adjacent property by, for example, ground movement off site or by the changing of ground water patterns, or by vibration, there may be grounds for an action for nuisance.

The case of *Rylands* v. *Fletcher* (1868) established the principle that, if a person brings on to his land any substance which is liable to do damage if it escapes, he is liable for all the damage which results if it does escape, whether or not the damage is reasonably foreseeable.

Control of Pollution Act 1974

The Control of Pollution Act allows occupiers or other persons affected by vibration causing damage, to apply to a magistrates' court for an order to abate the nuisance caused by such vibration. The municipal authority may also have taken powers to fix limits for machinery induced vibration, and it is the onus of a contractor to ascertain and comply with any such limits.

Damages resulting from vibration if proven, can give rise to a civil action for compensation. (For further information see *BRE Digest 353*, Watford 1990.)

Chapter 9
Customer care and quality assurance

Customer care starts with the interpretation of the customer's needs within the framework of the technical considerations associated with the work to be done. An understanding of the client's needs is necessary to ensure customer satisfaction. Satisfied customers are the best advertisement a firm can have. They will promote the company by word of mouth recommendation, which is undoubtedly the most effective method of selling products and services.

In underpinning, the 'customer' may in fact be a number of people: the engineer, the insurance company and the householder. Other authorities will also be involved.

- *The mortgagor* will be concerned about the effect of the damage on the value of the property.
- *The local authority* will be interested in health and safety matters and ensuring the work complies with the Building Regulations. The authority will also be concerned with potential nuisance and/or damage to neighbouring properties.
- *The public utilities companies* will be concerned about their connecting services (cables, pipes etc.) and any alterations required to these.
- *The contractor*, who is responsible for carrying out the final remedial and repair work, will want assurance that the householder can pay for the work and that the work can be carried out without causing further damage. The contractor will also be concerned with the serving of all notices and applications relating to the work.

Clearly, each of these parties will approach the task of rectifying the damage with different concerns and priorities. The underpinning contractor should be aware of the issues facing each of them. An appreciation of all the concerns will inevitably improve the perceived level of service offered.

Householders' concerns

Householders will become aware of subsidence, settlement and heave damage as they find doors beginning to stick, structural distortions and the appearance of cracks. Immediately they will worry about the safety of their homes and the potential fall in value of what, for most people, is their largest capital investment. Their apprehension will grow as they imagine the upheaval caused by work to rectify the damage and, worse, the cost of doing so. The perception of the money involved is often magnified by a lack of knowledge but, nevertheless, it is still very successful, conjuring up visions of a major drop in living standards.

Damage resulting from ground movement can be covered by insurance. In fact, building societies and banks lending monies for house purchases require the premises to be insured against subsidence and heave damage so as to provide security for the loan. Householders should be encouraged to seek advice as early as possible. For some, the local Citizens' Advice Bureau is often the first step. The legal profession, too, will provide help for a price. In all cases, the insurance policy must be read extremely carefully. It should cover the cost of repairing the damaged property to a habitable, serviceable and marketable condition. If this is uneconomical, a cash settlement to cover the loss in market value is an option. The policy will not cover improvements. It is essential therefore to ensure that the householder has correctly clarified the wording of the policy, particularly as regards inherent defects and possible lack of maintenance. To safeguard their position, householders need to make sure they have:

- complied with all of the 'small print', particularly the 'claims conditions' on prompt notification of damage in the correct and designated way laid down in the policy documents; and
- understood any exclusion clauses in the policy, such as how damage resulting from demolitions, structural alterations, lack of maintenance, faulty design and/or defective materials are dealt with.

In addition, if the owners wish to sell the property prior to rectification, they need to confirm that the insurance rights will be transferable to the purchaser. Professional advice should be sought if there is any doubt as to the extent of the cover.

Once the householder has lodged a claim, the insurance company will send a representative to confirm that the claim is valid (i.e. that subsidence or heave has occurred due to ground movement). At this stage, the householder might need to be made aware that investigations into the cause of damage – and whether the effects are ongoing or have ceased –

can take a long time. The insurance company may reserve the right to carry out long-term measurements before reaching a decision.

Insurer's viewpoint

The insurance company will be interested primarily in determining the cause of the damage and whether or not it is in fact an insured risk resulting from ground movement. There are also the questions of whether the amount for which the property is insured is adequate and whether there is a proper balance between the premiums and likely expenditure. In addition, the date of occurrence of the settlement or heave is pertinent where the period covered is defined in the policy.

In processing the claim, the insurer will appoint a 'loss adjuster' to deal with the property owner. This will usually be a specialist company with a background in building/engineering/insurance. The loss adjuster will act impartially within the terms of the policy, with a duty of fairness to both the insurers and the policy-holder, to investigate the claim and negotiate the settlement.

Handling underpinning projects

When dealing with an enquiry for underpinning works, the first task is to ensure you understand exactly what is needed. This is an essential step but it is not always an easy matter. General building contractors, experienced in their own work and products, may make unwarranted assumptions. Furthermore, householders may think they know what is needed and ask for it without knowing exactly what is possible or what might be more satisfactory for the purpose. It is, therefore, imperative at the outset that underpinning contractors scrutinise the enquiry or, where they are asked to design the scheme, insist on receiving the necessary information to do so. If anything is in doubt, or if there is information lacking, confirmation must be sought.

If assumptions have to be made they should be recorded and explained to the client(s). There is no sense in giving the client a solution that later proves to be unsatisfactory or inadequate or, conversely, over-designed. Good, value engineering must be the goal. Correct advice, with discussion early on, establishes confidence and ensures repeat business. Fulfilling the requirements to the satisfaction of all concerned is the route to successful customer care.

The objectives of any underpinning work must be:

❏ to restore and maintain stability to the property; and
❏ to ensure the satisfaction of the client.

Accordingly the work should be executed in a manner designed to ensure minimum disruption or nuisance. Obviously the choice of underpinning solution and equipment bears on this, particularly as regards noise reduction and tidyness of the site.

Clearing up rubbish as the work proceeds is a good start to ensuring a safe site. On completion of the works it should be the rule to leave everything in a habitable state as near as possible to the standard found before work commenced. Consideration should be given to reducing any annoyance to neighbouring premises and the public at large. It is useful in this respect to communicate with all parties concerned before work commences to explain the reasons for the work and the procedures and methods to be used. (A simple method statement is useful here.) In this way people feel involved from the beginning and may as a result not be so critical as the work proceeds. The absence of complaints often indicates satisfaction. Complaints that are registered should be dealt with promptly and viewed not as a nuisance but as an opportunity to bolster good customer relations.

Any problems with access and working space that cannot be overcome by the use of specially designed plant should be noted and dealt with by, for example, the removal of doors etc, with attendant assurances of adequate and proper restoration on completion.

Ensuring the safety of occupiers, visitors, neighbours and the public generally is at all times both a statutory duty and obvious good practice. Should it be necessary to excavate across footpaths or driveways, safe alternative routes must be provided. Water, gas, electricity and telephone services need to be maintained in a usable state. Where they have to be cut off for any reason, prior warning must be given and restoration should be made as soon as possible. In the case of accidental damage, prompt repairs must be a priority and an explanation should be given to the affected parties as soon as possible.

Restoration work to facade, gardens, drives etc. should not be overlooked. Any cracks, both structural and aesthetic, need to be made good using materials similar to the original. Restoration is important to the occupiers. Often in their minds, incomplete or bad redecoration can be synonymous with bad foundation repairs even though the repairs may be sound. Decoration is not only a necessity but is also an exercise in good customer relations. Clients' perceptions of the finishing and visible work often colours their view of the whole contract.

On completion of all works, someone should visit the premises within a few days of leaving as a follow-up to ensure the occupier is satisfied and nothing that should have been done has been overlooked. The question or otherwise of any guarantees is best dealt with as an office matter. A polite completion letter along with the invoice is always a good idea after the follow-up visit.

The challenge must be to bridge the gap between the firm and the client. Customer care may be director driven but really depends on the attitudes and behaviour of all the people in an organisation, particularly those actually carrying out the work. Service excellence and customer care is a continuous process and the biggest impact is made at the interface between workforce and householder. The work operators should be trained in customer relations. Two essential qualities are politeness and respect for occupiers.

Finally, remember that the reduction of nuisance ranks equally in customer service terms with successful completion of works. Consideration needs to be given to noise, access and working space. These are discussed in the following sections.

Noise

Underpinning operations involving piling and concreting are often carried out within premises using compressors, pistons and other noisy equipment. Levels of 100–140 decibels can occur. Such values are higher than work levels laid down in the Noise at Work Regulations 1989. Written guidelines entitled 'Noise Control and Hearing Conservation Means' have been published under the heading 'Noise in Construction' by the Construction Industry Advisory Committee (CONIAC, HMSO, London) and CIRIA Report 120.

Noise depends on the method and plant used, and measures such as reducing the air pressure required to drive pneumatic equipment, silencing exhausts, muffling and screening work on standing compressors by sound absorbing matting will all help.

It is always useful to explain to the householder and to adjacent owners, what you are going to do, its likely duration and the fact that all feasible noise reduction measures will be adopted. State, too, that work will not occur after normal working hours, say 16.30 (4.30 PM).

Access

Much underpinning work involves operating in places with poor access, resulting in the need for manhandling of heavy and awkward equipment and materials into and around domestic premises by workmen whose boots are likely to be dirty. It is therefore important that the workforce be trained to minimise dirt inside houses, to clean up after work each day and to leave the premises as habitable as practical. Politeness in all dealings with householders pays dividends in this respect, as do explanations of what has to be done. In addition, the access and site generally must be kept tidy and free from hazards likely to cause tripping or slipping.

Working space

Safe working must be the cardinal rule. Sufficient space to accommodate plant, equipment and workers should be provided. As far as possible the working space should be isolated so as not to interfere with the normal routine of the household. In this connection, specially designed low-headroom, compact plant is available. Specialist firms often design their own equipment to this end.

In the case of mass concrete underpinning, most of the work can and should be executed from outside the premises. It goes without saying that the sides of any excavation need to be adequately supported against collapse to ensure the safety not only of the property but of the occupiers and the workforce. Part of any excavation, particularly in traditional mass concrete underpinning, will be under the walls. Operatives should work adjacent to but not below any unsupported structure.

The working space should be tested regularly to ensure the atmosphere is not toxic, flammable or deficient in oxygen.

The location of public utility services, such as water supply, drains and sewers, gas and electricity and telephone wires, should be ascertained before commencing work. Care must be taken to avoid disrupting these services.

Access for occupiers and their safety

Provision must be made to ensure that occupiers have clear, safe and uninterrupted access to their household at all times. This can involve constructing temporary walkways and suitable guarding of excavations. All disruption or damage that results to doors, paths, decoration and so on should be reinstated or improved as part of the operation, to the satisfaction of the householder, before leaving the premises.

Quality

All of the issues already discussed are part of providing a quality service.

In this sense, quality is indicative of superior performance and customer care, and involves a commitment to service excellence. In summary, a contractor claiming to 'care for the customer' must take the following steps.

- Plan the work and inform the customer what to expect
- Minimise the potential nuisances, particularly noise
- Provide suitable access for the work and for the householder
- Maintain safe working spaces and use safe working practices
- Ensure the end result restores full structural safety
- Repair all cracks, including aesthetic ones, and redecorate competently
- Think *quality*
- Give *satisfaction*

Remember always to follow-up to ensure complete customer satisfaction as your reputation can be the best guarantee of repeat business.

The ultimate test of your quality will be a full structural survey carried out as part of a mortgage valuation. This will be used by the lender to decide whether to offer a loan, and how large, to a prospective purchaser. It will also provide an assessment of the efficacy of the repair work. If the valuer of a restored property reports no further defects or no risk of further damage, you can legitimately claim to have provided a total quality service.

Communication and quality

No apology is made for re-stating the obvious, namely that at all times both before and during an underpinning contract, the importance of clear, regular communications between all parties, along with meticulous recording of work, is essential for quality. This sounds simple but such procedures are often neglected during the course of a project.

Initially a pre-start meeting between designer and operator should take place to ensure that the method statement for executing the works is fully understood and capable of being adhered to, subject only to unforeseen circumstances. The client should then be advised in lay terms of what is to be done and what is to be achieved by doing it. As the work proceeds, communication will ensure that, should conditions change, costly construction delays will be avoided.

Quality control as part of quality assurance

Quality assurance provides a degree of overall confidence in materials and workmanship, but quality control on site is of particular importance in underpinning. Quality assurance begins with the site and ground investigations, influences design and persists until the job is completed. Quality control on site is the province of the individual operator and the supervisor.

The effectiveness of the quality of a given project depends on cash availability, the experience of the operators, the quality of the design based on site data, and the methodology adopted, as expressed in a method statement. Most engineers would point to inadequate site information (and, therefore, cash) as the main reason for potential failure; others would cite operator training as a potential weakness.

Consultants, contractors and building inspectors are now aware of the advantages and necessity of such quality assurance and quality control in all operations. 'Out of sight, out of mind' is a dangerous maxim in underpinning. Increased expertise in geotechnical investigation, increased levels of training, increased knowledge, experience and expertise all undoubtedly enhance ultimate quality standards and eliminate many sources of failure. Quality is cost effective in reducing recall and wastage.

Accreditation

In the UK, British Standard BS 5750 and the International Standards Organisation ISO 9000 series are the governing standards of quality, as is ISO 9000 for the rest of the European Union. The word 'quality' usually carries connotations of excellence, but in quality assurance quality is not necessarily indicative of special merit, even though the pursuit of excellence is the desirable objective. In the context of Standards, quality is used in its engineering sense to convey the idea of compliance with a defined or specified requirement of value engineering, fitness for purpose and above all the satisfaction of the client. This concept means that performance has to be measured and controlled, and the idea behind quality assurance is to involve everyone in improving all activities on a continuous basis to meet or exceed the expectations of clients.

Numerous third-party accreditation bodies exist to inspect and assess quality performance and quality assurance and to issue registration once acceptable standards have been achieved in terms of the various standards against which the company is being assessed. Such registration continues for as long as the company maintains records acceptable to the certified body and is capable of tracing back all activities to origins. Thus a company with accreditation can provide evidence of quality assurance

Table 9.1 European quality standards

Standards organisation	Quality assurance system
International Standards (ISO)	ISO 9001–9003
European (CEN)	EN 29001–29003
UK (BSI)	BS 5750: Parts 1 and 2
Germany (DIN)	ISO 9001–9003
France (NF)	X50–131 6 & 50–133

to a client, demonstrating that they operate to accepted standards and can therefore be relied upon. The main European quality system standards are as given in Table 9.1.

The cooperation between members of the EU on quality assurance has resulted in a consolidation of various national standards to provide a system of accreditation and inspectors to provide certificates under the nomenclature BS 5750/ISO 9000/EN 9000. Thus each participating county requires similar documentation, comprising a quality manual describing the system and evidence of record-keeping, aimed at ensuring traceability and an ongoing commitment to quality.

Chapter 10
Prevention is better than cure

The UK National House-Building Council publishes a booklet entitled *Preventing Foundation Failure in New Dwellings* (latest edition, 1991). The work was intended primarily for registered house builders, architects and engineers operating in the private housing sector, but it also provides valuable guidance for anyone concerned with reducing the incidence of foundation failure.

Approximately 60% of major building defects can be attributed to foundation/ground movement. The main problems are bad infilling, bad workmanship, subsidence, settlement and heave, and the selection of unsuitable sites comprised of hazardous or unsuitable ground.

The subject of prevention can be considered under the following headings, corresponding to the stages involved before building work commences.

- The preliminary investigation and initial selection of any building site.
- Identifying the need for abnormal or more extensive site and ground investigations.
- Engaging the services of a competent engineer with relevant qualifications and experience.
- The use of specialist foundation engineers/geotechnical engineers.
- The careful design of the foundations themselves.

Recommendations relating to each of these stages are outlined in the next section.

Good practice for prevention

The preliminary investigation can yield valuable information about the suitability of a site and should be carried out thoroughly. Initial reconnaissance should as a minimum cover the following points.

- Establishing, as far as possible, the history of the site.
- A cursory inspection of any neighbouring structures to ascertain if there have been any structural problems.
- A discussion with local inhabitants, and with local and public utility authorities.
- A study of trial pits to determine the general ground substructure down to depths below normal foundation levels.
- An inspection of geological and ordnance survey maps and any available archival records.
- Obtaining opinions on the state of the ground in both winter and summer.
- Logging the positions of tree removals, with a view to determining the effect on the water table.

Chapter 3 gives more details on the procedures involved in site investigation. This preliminary study should identify any potential problems with the site and enable you to assess the likelihood of any abnormal foundation requirements. This will allow you to 'guesstimate' the likely additional costs.

The next step will be to conduct a more thorough, detailed ground investigation. This could involve:

- A dimensional and annotated ground profile, noting the water table at the time of the survey.
- The location of trial pits and/or borings marked on the proposed layout plan with levels noted. (See the section on ground investigation in Chapter 3.)
- A full description of the strata encountered.

If the ground investigation reveals any particularly difficult conditions, you should consider employing the services of a competent engineer, a consultant, or a specialist contractor. In selecting this type of service you should consider the following points.

- Define clearly what you require from them in an 'information brief', providing what data you have already collected and specifying a time-scale if possible.
- Screen the list of suitable suppliers by personal enquiry and seeking recommendations. Cut it down to, say, two and ask them to quote you. Then choose the one with the best experience of ground and foundation work or the best qualifications.

Once you have a comprehensive description of the ground conditions, the next task is to select the optimum type of foundation for the site. Table 10.1 shows roughly what types of foundation are appropriate to some general soil conditions.

Table 10.1 Selection of type of foundations most suitable for given ground conditions

Foundation	Ground condition			
	Peat, loose sand or gravel	Shrinkable clays	High water content ground	Filled or made ground
Strip foundations	Not recommended unless action with foundation provides sufficient depth or strength to offset low bearing capacity of ground	Not recommended	Not usual unless special precautions are taken	Not usual without prior ground treatment by vibrocompaction or equivalent
Pad and beam	Recommended	Not recommended	Can be used	Only with very shallow fills
Piling	Best option	Suitable if sleeved or with slip membrane	Suitable at depth	Best option
Raft	Possible if piling not available	Not recommended	Suitable	Recommended if fill is reasonably stable
Ground treatment	Possible	Possible but unlikely	De-watering by various means	Possible
Specially designed system (pile and precast beam, often patented).	◄——————— Can be used universally ———————►			

The next step is to take the final decision on the site layout, the type(s) of dwelling(s) to be erected and, particularly, the foundation design. Advice on building layout, particularly the location of walls, is needed along with a professionally designed foundation layout.

In raising capital or mortgages, evidence that a comprehensive study has been undertaken may be required. It is advisable therefore to obtain a written report and certification of the foundation design stating that the substructure work has been designed after an adequate site and ground investigation, in line with relevant Codes of Practice and Standards. Relevant Standards in the UK are:

BS 648   Weights of building materials
BS 5930   Site investigation

BS 6399 Loadings (Dead and imposed)
BS 8004 Code of Practice
BS 8110 Code of Practice
BS 5837 Trees in relation to construction

In addition, general good and accepted practice is laid down in the Building Regulations.

Once the work has been approved by a regulating body, adequate supervision of construction is necessary to ensure adherence to the drawings and specification and to the quality of work and materials.

Figure 10.1 summarises the steps that should have been taken before work commences on a new building project. If adequate data on the ground conditions are available and if appropriate design and building expertise are employed, it should be possible to prevent any risk of foundation failure.

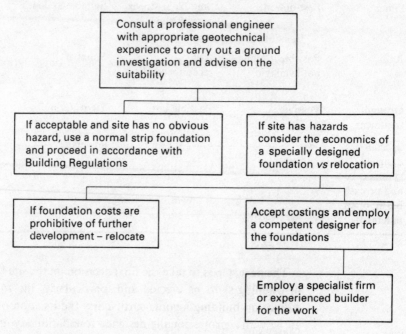

Fig. 10.1 Summary of steps to prevent foundation failure

Hazards and precautions

The substantial costs of remedial work necessitated by ground movement make it not only desirable but also cost effective to pre-plan construction work to the highest standards. Prevention hinges on having comprehensive data on the site and expert interpretation to identify potential hazards. Possession of this knowledge will facilitate the design of foundations that will withstand the stresses from ground movement in service. The following sections describe some common hazards and appropriate precautions.

Firm shrinkable clays

Soils such as these are prone to both settlement and heave, depending on the seasonal movement of the groundwater. As the water content decreases in summer, the soil will shrink. With the ingress of water during the winter months, the clay expands and causes heave. Some clays can experience movements of 20–40 mm in both directions. The same problem can occur in situations of water drainage or where there are leaking sewers or defective water mains. The removal of vegetation, particularly trees, also severely alters the water content of the ground.

Sensible precautions

- Design the layout and position of the building(s) to avoid known hazards as far as possible, retaining hazardous areas as gardens and open spaces.
- Use professional advice to design the foundations to accommodate the type of movements anticipated, particularly heave. Shrinkage and heave can be assessed by referring to the actual clay content of the subsoil and its plasticity index as shown in Table 10.2.

Table 10.2 Typical ranges of shrinkage or heave in clays

Type of clay	Plasticity index (%)	Clay content (%)	Shrinkage or heave (mm)
Kimmeridge, Gault and Reading clays	54–70	59–70	Very high (>40)
London, Oxford clays	28–50	56–65	High (30–40)
Clayey silts	11	19	Low (<25)

Silts and peaty ground

These soils have low load bearing capacities and settle under load. Where the thicknesses of these strata vary across a site, differential settlement can occur. Also, peat can readily lose its water by drainage or abstraction.

These effects can occur when the layers are near the surface or down to about 5 m below the surface. In the latter case the situation may not be shown by trial holes. If peat is located by examination of geological maps or from local information, the depth of trial excavations or bore holes should be extended down to at least 5 m.

Precautions
Sensible precautions are similar to those for clays. There is also the possibility of excavating the peat or silt and replacing it with good backfill, and then providing a properly designed raft foundation of suitable thickness and depth below all walls. If the peat/silt is at a depth greater than, say, 1.5–2 m, it will become necessary to pile through the soft layers to good bearing ground. If there is no good bearing ground at an acceptable depth, it is possible to design a stiff box raft. If there is an acceptable surface layer of at least 4 m above the peat or silt, a reinforced strip foundation is a possibility.

In all cases on peat and silts it is advisable to provide movement joints and flexible drain jointings, and reduce loadings by adopting lightweight construction. Consider also soil stabilisation.

Mining subsidence

This can arise in areas of coal mining, salt mining and mineral workings. The most common problems are in coal mining districts. Fault lines in such areas constitute serious hazards and building on them must be avoided. Before work is contemplated in areas likely to be affected by subsidence, the local authority and the mine owners must be consulted.

Where the existence of unrecorded abandoned workings is suspected, further local enquiry is warranted and a full ground investigation, supervised by a mining engineer, should be undertaken. The advice of a mining engineer and the mining authorities on the type of foundation is essential and the incorporation of a jacking system in the foundation design may be required where subsidence is ongoing and the building itself warrants it.

Installed jacking system for mining subsidence areas
During the initial construction of a dwelling in mining areas it is feasible to install a jacking system with sensors which determine and monitor any differential movement. At a predetermined level of movement, the jacks can be activated (raised or lowered) to accommodate either settlement or heave. It is costly to install and relies on the efficiency of the sensor mounting system and effective jack maintenance.

Ground improvement techniques

The bearing capacity of the ground is obviously a crucial factor in the design and integrity of foundations. Techniques have been developed to strengthen the supporting strata and improve their physical properties. These range from the replacement of underlying soil, through processes to increase soil densities in situ, to moisture control systems. Some key methods of improving ground conditions are outlined below.

Soft clay sites

To overcome the problems associated with clay soils, it can be economically worth while to remove up to 2 m of the topsoil and replace it with well-compacted crusher run stone filling. A raft can then be formed on top of the fill. In the case of strip foundations on compacted fill trenches, the fill must be two and a quarter times wider than the footing or two times deeper than the foundation width.

Rehydration

A new technique suitable for clay soils has been developed by London based consultants Packman Lucas Associates. The system uses permanently installed probes or well points connected to a water supply to rehydrate the ground and maintain an equilibrium moisture content (see Fig. 10.2). This will all but eliminate the movement caused by variations of moisture levels in shrinkable clays.

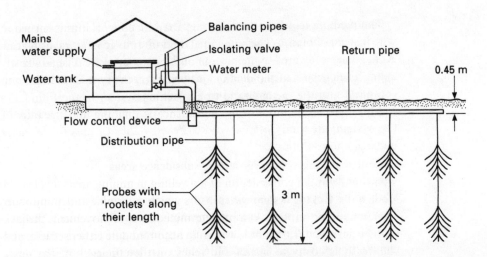

Fig. 10.2 The Packman Lucas moisture management system

Packman Lucas has designed three types of probes to carry out the rehydration to low-, medium- and high-capacity levels. The first two simulate a reverse tree root system. In the low-capacity version a 100 mm hole is drilled into the ground to an appropriate depth, approximately 3 m. A special sleeve carried by a piling rig is lowered into the hole and two 800 mm long blades are forced out into the soil. They are then withdrawn, the device rotated and the process repeated four times up to foundation level. Sand is then washed down into the cuts produced by the blades and a water pipe is embedded therein. The top part of the hole is concreted with the water pipe projecting and coupled to the water supply (see Fig. 10.3).

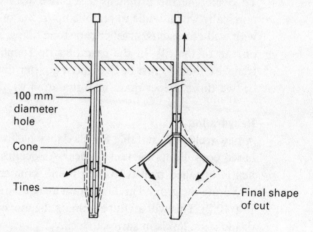

Fig. 10.3 Installation of a rehydration probe

The medium-capacity version is similar but uses 16 strands of piano wire instead of blades. The high-capacity probe has four 'tines' or blades. As the cone is lowered to the bottom of the hole, tines are pushed slightly apart. The probe is then pulled out leaving a large cut, which is filled with compacted sand and connected to a water supply.

Although careful monitoring of moisture content is a major feature of this system, the total outlay is said to be only 25% of the cost of many underpinning solutions.

Other methods

With some clays it is possible to stabilise the ground by injecting lime. This method is normally used in conjunction with excavation to remove the top layers and replace them with imported soil that is stable and compact or pulverised fuel ash. It is an expensive option but sometimes offers the best solution.

Cohesive clay soils can be improved by vibroflotation. This technique is outlined in the following section on sand and gravel sites.

Curbing the spread of tree roots will also prevent moisture content problems in clay soils. Methods of doing so are described later.

Sand and gravel sites

Sands and gravels usually contain a great number of voids and are therefore prone to gradual and differential settlement under the load imposed by a building. The bearing capacity of such soils can be improved by compaction. The three chief methods of improving loose soils are:

(1) *Mechanical*
- Vibratory (vibrocompaction and vibroflotation)
- Preloading
- Dynamic compaction
- Water removal
- Ground replacement

(2) *Physico-chemical*
- Electro-osmosis
- Freezing

(3) *Chemical*
- Grouting.

Where the sands or gravels are deep and do not have the required bearing capacity, it is most common to attempt soil improvement using the vibratory or dynamic compaction methods.

Vibratory compaction

Vibrocompaction makes use of a heavy vibrating roller to settle the underlying ground, thereby increasing its density and improving the load bearing and deformation characteristics. The two methods of vibroflotation are:

(1) sand compaction, used to densify non-cohesive free draining soils; and
(2) the stone columns technique for use in cohesive clay soils.

The sand compaction method   A large torpedo-shaped vibrating poker is suspended by a crane or similar device above the selected point on the ground. By means of its own weight, with assistance from water jets located at the bottom, the poker penetrates to the desired depth, whereupon these water jets are switched off. The water flow is then

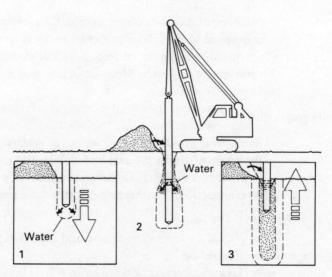

Fig. 10.4 The sand compaction method of vibroflotation

transferred to upper jets and vibration occurs. Sand or fine gravel is fed into the top of the drilling to compensate for the compaction occurring. The poker is raised incrementally to form a solid cylinder of sand about 3 m in diameter. Similar operations are carried out across the site so that the compacted areas slightly overlap and a significant deep area of reasonably uniform enhanced density is formed. (See Fig. 10.4.)

The stone column method This technique also uses a vibrating torpedo-shaped poker. It penetrates into the soil by the combined effect of its own weight and the jetting action of water or air. In the case of air jetting, the in situ soil is displaced and compacted laterally, whereas with water jetting some material is flushed out to the surface.

The poker is withdrawn once the desired depth is attained and stone is tipped into the hole in 0.5 m increments; the poker is then repositioned and the stone vibrated, forcing it downwards and outwards into the walls of the hole. The process is repeated in 0.5 m steps until the surface is

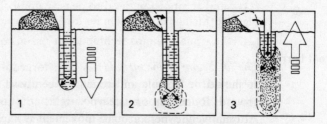

Fig. 10.5 Vibroflotation by the stone columns method

reached and a dense column of granular material interlocked with the in situ soil is formed. (See Fig. 10.5.) Repeat columns at predetermined centres to produce an area of enhanced bearing capacity.

Dynamic compaction

A heavy steel hammer or tamper is dropped into the ground from a height to achieve compounding. The resulting high energy impact transmits shock waves through the ground to the depth to be treated (governed by the height of drop and number of blows). This reduces the voids in the soil and causes consolidation. The treatment is carried out over a grid pattern to ensure overlapping areas of compaction. (See Fig. 10.6.)

In all cases of vibration or dynamic compaction, in situ tests are needed to validate expected results.

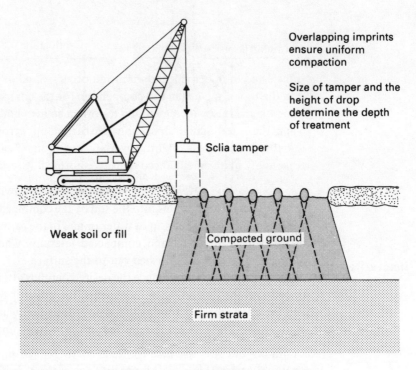

Fig. 10.6 Method of operation for dynamic compaction

Tree root barriers

It should be feasible and practical to restrict the growth of tree roots towards foundations by interposing a barrier of steel, concrete or grout. Alternatively, where ground movement is likely to be a problem, reinforced polythene sheeting can be used. Where this type of flexible barrier is used it should be installed in a trench dug to an appropriate depth (dictated by tree root depths) and situated between the premises and the tree. The polythene is placed in contact with the excavation side nearest the tree sources and back-filled to hold it in place. Care is needed to ensure the back-fill does not damage the flexible barrier and it should therefore be free of stones. Any joints should be well lapped (at least 450 mm) and preferably sealed with adhesive tape.

The barrier should extend vertically to below the root depth by at least 0.5 m and upwards to ground level to obviate root bridging. The system (illustrated in Fig. 10.7) is likely to prove uneconomic where required depths exceed 4–5 m. (Remember that some tree roots penetrate as deep as 6 m.)

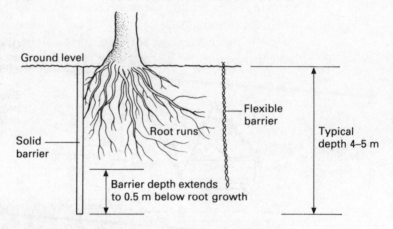

Fig. 10.7 Using barriers to control the growth of tree roots

Innovative foundation systems

In the battle to prevent failures, innovation has a part to play. Innovative foundation designs are, in essence, modifications, adaptations or extensions of existing systems. They are tailored to the ground conditions but are specifically designed to reduce both construction times and costs. This makes it possible to specify foundations that will, hopefully, eliminate the need for future remedial and repair work in situations where they would not otherwise have been economically viable.

Innovations around at present are frequently configurations of piled systems using jointed precast concrete beams in variations of the pile and beam underpinning designs used after ground failure has occurred. Some typical examples are described in the following sections.

Precast grade system

Shafts are drilled to a designed pattern and depth around the perimeter of the premises and along the lines of load bearing internal walls. Short plastic formers are inserted into the top of the shafts, and link up with the underside of the proposed foundation beams.

Precast reinforced concrete beams are then set out on blocks to conform to the configuration of the proposed building. These 'grade' beams are levelled up by means of timber wedges to the desired foundation level. Pre-drilled holes in the beam are designed to coincide with the tubed shafts. Through these holes are inserted rebar type steel rods, as specified by the design, which hang from the surface of the beam by a hooked end of at least 40 diameters extent. Concrete of $30\,\text{N/mm}^2$ strength is then fed into the shaft up to the level of the grade beam and within the formers. After 24 hours the wedges are removed and a ready-for-use foundation exists. The gap between ground level and the beam can either be filled with concrete or left. (See Fig. 10.8.)

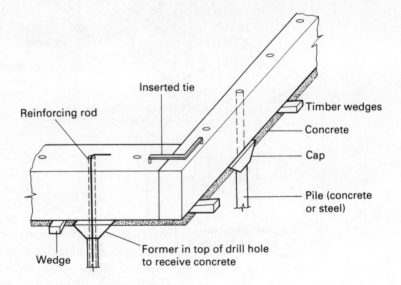

Fig. 10.8 Precast grade beams

A designed foundation system

This innovation involves the design, supply and installation of a complete fast track foundation system for houses on all types of ground. It uses piles or cones carrying reinforced concrete beams.

One such recent innovation, which won the 1995 National Construction Innovation Award, is a modular foundation system as produced by Roger Bullivant Limited. It consists of piled supports, precast concrete foundation beams and concrete waffle flooring cast using pumped microconcrete with a high pulverised fuel ash content around polystyrene formers.

The principle is shown in Figs. 10.9, 10.10 and 10.11. The system is designed to be as flexible as possible, utilising any type of piling, depending on the ground conditions, and using a range of precast beam sections. It was designed for poor ground conditions where contractors

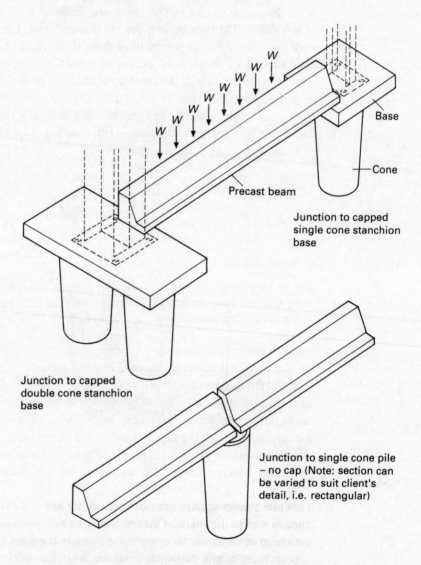

Fig. 10.9 Precast concrete beam on cone pile system

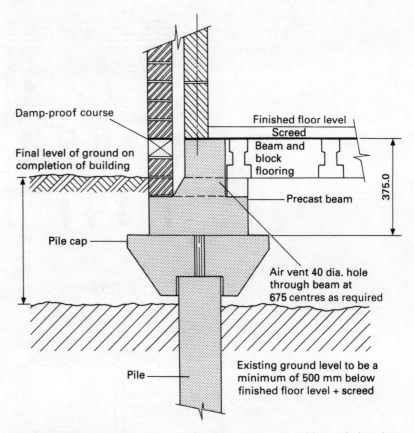

Fig. 10.10 Precast tee-beam is a reinforced concrete ground beam designed to carry wall and floor loads in traditional low rise construction. Also useful in single storey shed construction

want to minimise excavations, but the company recommends the system as a cost-effective foundation even in good ground.

Bullivant has developed its own plant and equipment to mechanise the construction process as far as possible. The entire system can be built using only two pieces of machinery, both of which can be delivered to site on one low loader. The first is a small skid steer loader. The second is a crawler-mounted piling rig, capable of installing augered, bored or driven piles, which converts into a crane to lift the precast elements into position.

While conventional piles can be used, a preferred method is to use a precast unit in the shape of the frustrum of a hollow cone and, depending on the ground, either to drive a pile through the cone and concrete the top to form a pile cap (pile, connection and cone) or, if the ground conditions permit, merely to concrete up the cone itself. Cones can be

140 *Underpinning*

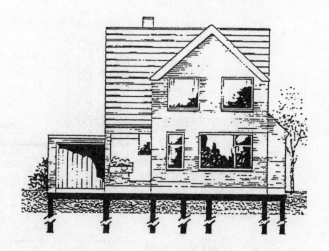

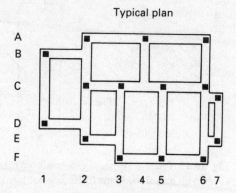

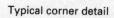

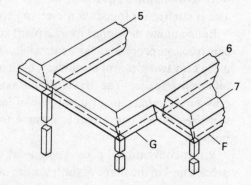

Fig. 10.11 Tee-beams on piles

produced in several lengths to ensure adequacy of bearing capacities. The cone shaft can be constructed either by precasting or by the use of a former mandrel and casting in situ. Precast concrete reinforced beams are then set on the cone supports. These beams will vary in dimensions and reinforcement depending on the design and may be rectangular or of inverted type shape.

Normal domestic and light industrial loading from 30–100 kN/m of foundation run can be catered for, with total flexibility of beam reinforcement and pile centres, so providing economic layouts and controlled beam spans.

In a designed foundation system solution, there are three distinct operations:

(1) design;
(2) pile or cone installation – either augered, bored or driven to the design layout; and
(3) beam installation – set level and joined together by in situ concreted reinforcement.

Steel helical anchors

Useful for low rise, lightly loaded structures, steel helical anchors are rebars or lengths of square or circular solid steel rod with one or two helixes welded to them at right angles to the vertical axis. The anchor is rotated into the ground to the desired depth by mechanical means, such as a portable drill rig, or, in the case of rebars, driven. An extension piece can be bolted or welded on for greater depths. (See Fig. 10.12.)

The helix provides an additional bearing surface as does the rebar, to facilitate safe transference of the load to the ground. In the case of a

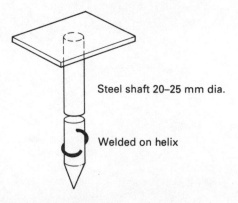

Fig. 10.12 Steel anchor with helix welded on

142 Underpinning

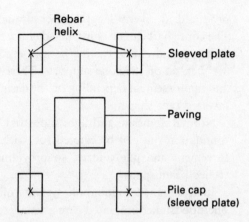

Fig. 10.13 Foundation configuration with steel anchors

heave situation, the helix provides resistance to uplift. If below the clay area, the helix itself is usually adequate. A purpose made cap is bolted to the top of the finished anchor and a concrete beam cast or placed on top to form the foundation, spanning between anchors. (See Fig. 10.13.)

Appendix A
Case studies

The case studies presented here are underpinning projects carried out by Roger Bullivant Limited. The solutions described demonstrate how the techniques discussed in this book have been used in practice. Including both underpinning and foundation solutions, the examples cover:

(1) Pile and needle underpinning
(2) An external solution using cantilevered reinforced concrete beams on piles
(3) The use of reinforced concrete beams and steel cased jacked piles
(4) A pier and knuckle solution
(5) A pad and beam solution
(6) The use of mini piles and jacking
(7) Stabilisation with angle piling
(8) A traditional underpinning solution
(9) The use of a jack piled raft
(10) Piled foundations for an industrial site
(11) Piled foundations on an open site.

Case study 1
Pile and needle solution

The inadvertent removal of some timber piles beneath two stone columns by a thrust boring contractor led to some structural distress in a courthouse building in Eire. Thrust boring was ceased immediately as the problem became apparent and the stone arches above were quickly shored up. A specialist contractor was called in by the scheme's consulting engineers to propose a solution.

Design

Not only was there a 1050 mm diameter thrust bore pile under the stone columns, but the main column was immediately adjacent to an existing pear shaped 1000 mm 'live' brick culvert. The design, therefore, took the form of a number of interconnecting concrete surrounded universal column section needle beams, sitting on spreader pads or beams, which in turn are supported on 165 mm diameter segmentally hollow stem augered mini piles. The total load of the main L-shaped pier is 1200 kN. This load is distributed on to 12 No piles, rating the piles at 100 kN each.

On-site works

Clearly the operation of installing the piles so close to the brick culvert and the connecting thrust bore pipe was delicate and had to be executed without vibration to either the structure or the culverts. It was decided, therefore, to use hollow stem augering methods to eliminate vibration and facilitate the pile installation through approximately 2 m of soft clays overlying dense sands and gravels. Restricted access and height required the use of Bullivant mini drilling rigs with 1 m length segmented augers. Having installed the piles to the close tolerances, reinforced concrete pads and spreader beams were constructed on both sides of the piers. A phased operation to install the universal column section then took place, with load transfer being achieved by the use of Armourex high-strength grout. The beams were then surrounded in concrete and reinstated.

Result

The stress induced by the removal of the original timber piles was reduced and minor repairs to the stone arches were carried out. The structure was rendered safe and stable.

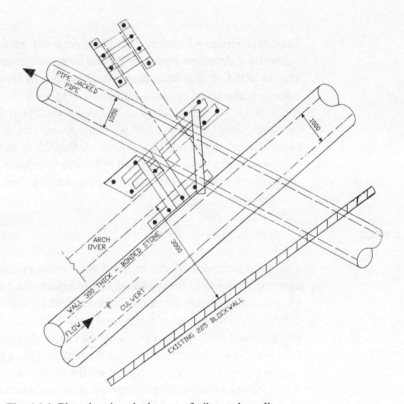

Fig. A1.1 Plan showing the layout of piles and needles

Case study 2
External solution using cantilevered reinforced concrete beams on piles

The initial enquiry from the consulting engineer, relating to the stabilisation of a detached house in Wolverhampton, requested an external only solution so that the clients could remain in residence. A detailed layout of the property was supplied along with the general builder's specification for the required remedial works. The property was visited to assess means of access and working space, and to obtain any local knowledge of the ground conditions. A neighbour recalled that the site had once been a sand quarry and in the rear garden a lace of weathered sandstone was noted on the rear steps boundary line.

Design

The underpinning system proposed was to use cantilevered reinforced concrete beams cut into the substructure brickwork and supported on 2 No 100 mm diameter cased piles (see Fig. A2.1).

From the knowledge of the ground conditions (i.e. the possibility of it being a filled site), cased piles would be required, with the compression piles being a normal driven pile but the tension pile requiring location at the toe into the underlying original sandstone/dense sand. This location was achieved using rotary percussive drilling techniques to form a socket at the base of the tension pile casing.

On-site works

The existing service pipes and cables were located and marked by the general builder prior to arrival on site. The following procedures were then carried out.

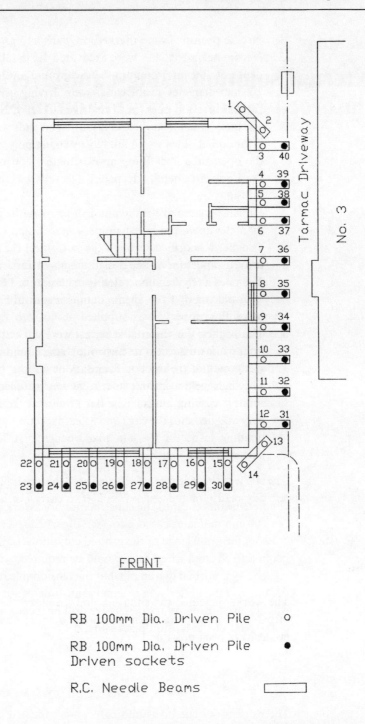

Fig. A2.1 Sketch showing layout of compression and tension piles for cantilevered reinforced concrete beams

1. Needle positions were marked on walls and ground.
2. Needle beam trenches were excavated 1.2 m, 300 mm wide and to a depth of 400 mm minimum below ground level.
3. 100 mm diameter steel casings were driving using a 95 Grundomat for compression piles (bottom driven to set of 10 mm in 10 seconds).
4. 100 mm diameter steel casings with open ends were driven using a 95 Grundomat. This was achieved by fixing polythene over the end of the open tube and filling approximately 500 mm of tube with dry sand/gravel/cement (a drypack). The tube was then bottom driven to refusal.
5. Utilising a standard Bullivant drilling rig with air flush and a 75 mm diameter down-the-hole hammer (on sectional 1000 mm long stems), a socket was drilled through the bottom of the casing into the then proved sandstone to the design length of 1500 mm.
6. The piles were then concreted using the mixed C35 N/mm^2 concrete after placing of 1 No 12 mm diameter central high-yield reinforcing bar with spacers.
7. The pockets for the needle beams were cut out of the substructure brickwork on a 'hit one, miss one' basis to minimise the temporary weakening of the existing foundations.
8. The high-yield reinforcement cage was installed in the trench with suitable spacing blocks. The bar projecting from the piles was bent over and tied into the cage and after inspection by the local authority building inspector the site-mixed concrete (C35 N/mm^2 mix) was placed using a vibrating poker.

The site was then cleared of all excavated material, plant and casings and left tidy ready for the general builder to carry out his surface reinstatement work.

Result

The works remedied the situation and stabilised the building. If access into the premises had been permitted a more cost-effective solution might have been possible.

Case study 3
Stabilisation using reinforced concrete beams and steel cased jacked piles

Roger Bullivant (East Anglia) was asked to submit a tender for stabilising the tower of a fifteenth century church in South Somercotes, located on the Lincolnshire fen. The tower had been showing signs of differential settlement relative to the main body of the church and this had resulted in the displacement of the keystones in adjacent arches.

Design

The design solution submitted was to excavate down and expose the tower foundations, cast reinforced concrete beams on both sides of the exposed walls after having grouted the rubble filled cavity in between. On completion of the beams they were to be connected in tension using stainless steel macalloy bars, sandwiching the original walls. Following this operation it was intended to drive 220 mm diameter steel cased piles through pre-formed holes in the reinforced concrete beams, connecting them to the reinforced beams via a system of anchor bolts and 25 mm thick stainless steel plates.

As the tower was actively moving and the substrata consisted of peat and silt above soft clay at a considerable depth, Bullivant recommended a 175 mm heavy-walled pile which could be jacked in to achieve the required capacity and eliminate any possible detrimental effects associated with a driven pile.

This system was accepted and piles duly installed, reaching a depth of up to 26 m below ground level. On completion of the piling, pre-determined loads were jacked into the piles to transfer support for the tower from the ground to the piles.

150   *Beams and jacked piles*

Result

The work arrested further movement and allowed the remedial works to arches and masonry to be completed.

Case study 4
Pier and knuckle solution

The property in question, situated in Minster, Sheppey, was investigated for subsidence damage by engineers Derek Crawley & Associates. They determined that the damage was occurring differentially to the flank wall of the property because of shrinkage of the underlying claygate beds strata caused by root activity of the adjacent hornbeam hedge. The request was to provide, if possible, a solution to limit the works to the external confines of the property at the flank wall area, together with returns to the front and rear of approximately 6 m at each location. The distance to the site boundary local to the flank wall was limited to 0.5 m and a proviso was put on the scheme to remove the hedge in this vicinity to enable the work to be implemented.

Design

Since the original construction was a narrow strip foundation founded at 1.2 m below ground level within the claygate beds (a clayey, silty, fine sand), it was decided to utilise a system comprising piers down the face of the original footings and reinforced concrete knuckles above the footing level to support the structure. To provide additional stability, it was decided to provide diagonal needles at the corners of the flank wall and tie all the piers and needles with a tie beam construction along the face of the existing wall.

The types of pile considered were a driven 220 mm steel cased pile and a 250 mm augered cast in situ pile. Either type was considered to be acceptable. The claygate beds had only a modest clayey content, although sufficient to be affected by the action of roots, so no form of anti-heave precaution was deemed to be required, taking into account that the remedial works were a partial scheme and did not cater for any internal walls that connected directly to the external wall in the vicinity of the underpin.

As a precautionary measure, it was decided to provide a 2 m length of slip sleeve to the upper level of the pile, either by pre-augering and driving the steel cased pile through the 250 mm diameter sleeve or, similarly, providing the sleeve to the augered pile construction.

On-site works

The works commenced initially with driven piles but this resulted in a shallow pile and severe vibration problems. This method was, therefore, abandoned and 250 mm diameter augered piles to a depth of 6 m were installed. The claygate beds contained more sand at greater depths but an open bore situation occurred with a firm base and, thus, the piles were constructed with the 2 metre sleeve and the T25 reinforcing bar installed to the design locations. The reinforced concrete works were then constructed at a suitable depth below existing ground level to support the structure as required.

Result

The project was proven to be more economical than a traditional scheme because of the time-scale and material costs. To have undertaken the works with the hedge intact would have necessitated an internal underpin which would have had severe repercussions in terms of disruption as well as increased costs.

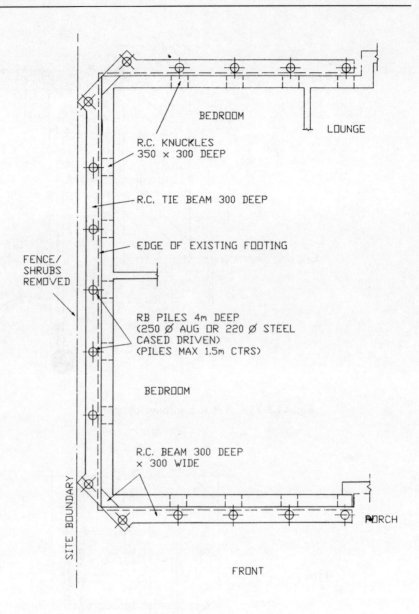

Fig. A4.1 Part plan of the pier and knuckle system

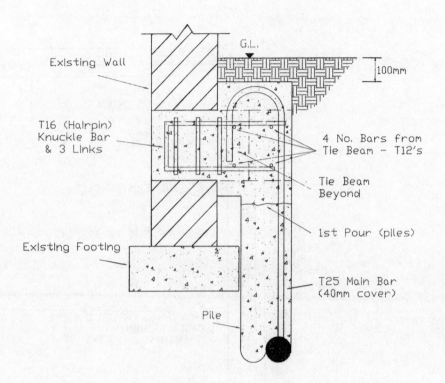

Fig. A4.2 Typical section through the pier

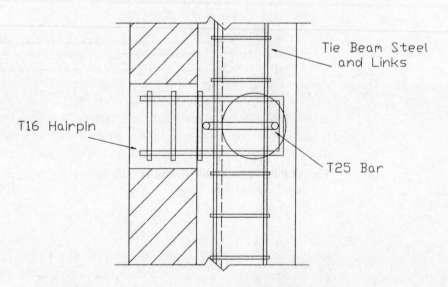

Fig. A4.3 Plan at the top of the reinforced concrete knuckle

Case study 5
Pad and beam solution

A specialist contractor was engaged to undertake underpinning works to a detached dwelling at Castle Bytham under the supervision of consulting civil and structural engineers. The property had suffered distress as a result of being founded partly on limestone bedrock and partly on what was thought to be a backfilled quarry.

Design

The pad and beam method of construction was proposed as it would give better resistance to any possible lateral movement of the fill materials than a mini piled solution.

The scheme involved the construction of 9 No pad bases founded at a depth of 2–3 m below ground level. The span between the bases depended on wall loading but averaged 3 m with pads being positioned to maximise the use of a standard beam size. The reinforced concrete (RC) beam was designed and constructed on top of the existing concrete foundation within the depth of the substructure of the building.

The schedule of proposed works included some initial investigation to assess the adequacy of the extent of the works. This was undertaken within the excavation for three pad bases at the limits of the underpinning works. As a result of this requirement the remaining pad bases were then constructed. It would be more usual for the RC beam to be installed as the initial operation in a contract of this nature as this would reduce the probability of further movement during the execution of the works.

On-site works

The pad bases were excavated by hand through clay fill material with limestone inclusions. They were extended up to a maximum of 3 m below ground level and 'tied' into weathered limestone. Four of the bases were cast with rebates, to allow the installation of hydraulic jacks in order that the superstructure could be jacked back to level to minimise the amount of rebuilding required at the major fractures.

Construction of the RC beams was completed in four stages with no two adjacent spans being removed at the same time. The existing substructure was broken out and 'cast in' adjustable stools were introduced at 600–1000 mm centres, depending on the quality of the superstructure brickwork and the load involved. Beam reinforcement was introduced to pass either side of these stools and the beams concreted using ready mixed concrete placed by hand.

Following a suitable period of curing of the concrete works, assessed by testing of concrete cubes, the superstructure was jacked to level with hydraulic jacks placed between the mass concrete pads and RC beams. The resulting void between the pads and beams was temporarily shimmed using steel plates, with the void being pressure grouted on removal of jacks to ensure adequate load transfer.

Result

The underpinning works together with removal and reinstatement of fixtures and fittings, superstructure repair and redecoration were undertaken in 6 months. The work remedied the problem and proved cost effective when compared with alternatives that were considered

Case study 6
The use of mini piles and jacking

Roger Bullivant acquired 12 detached houses – nine three-bedroom and three four-bedroom properties – in March 1991. The houses are of traditional design and construction. All had been constructed using reinforced concrete rafts with integral ground beams, on varying depths of fill. However, ground water ingress had caused localised softening of the fill material which, coupled with areas of lesser quality upper level fill, had caused differential settlement of the structures.

The raft construction provided the necessary degree of stiffness such that, on tilting, the structures flexed, producing nominal cracking rather than full structural fractures.

Design

A scheme was formulated to remedy the problem based on the installation of mini piles, followed by jacking to level of the existing rafts and culminating in complete refurbishment and restoration. Full approval was received from the National House-Building Council and Dudley Metropolitan Borough Council.

On-site works

An average of 20 No 220 mm diameter mini piles were installed to depths of approximately 12 m below foundation level for each house.

Structural steel brackets were fixed to the existing ground beams at pile positions to facilitate the jacking operation. Upon completion of the hydraulic jacking operation, permanent jacks were installed and concrete pile caps constructed.

The maximum height of jacking was 200 mm. Having arrested the movement and reinstated the structures to line and level, complete refurbishment took place, comprising structural repairs, new drainage, new service connections, redecoration, new driveways and paths and re-landscaping.

Result

The whole operation took just six months from commencement to completion. Disruption was kept to a minimum and the houses are now re-occupied by new owners who are benefiting from the full National House-Building Council 10 year Building Mark Warranty.

Case study 7
Stabilisation with angle piling

Tower 4 stands proud of York city walls and is built on an embankment of medieval fill. The tower had been subsiding for a number of years and previously efforts had been made to reduce the load on the embankment by replacing part of the core of the tower with polystyrene blocks. Settlement had, however, continued and it was therefore decided to stabilise the tower by introducing a series of mini piles. Bullivant won the contract after competitive tender and assessment of the proposed scheme by York city engineers.

Design

Ten piles of nominal 150 mm diameter and SWL 100 kN were required to be installed through the existing masonry and rubble core, then penetrating the medieval fill before founding in the underlying dense sands and gravels.

An angle pile system was selected as technically the best and most cost-effective solution. Problems anticipated were:

(1) The random rubble core of the walls collapsing into the bored hole.
(2) Obstructions within the medieval fill should driven piles be introduced once pre-boring of the walls had taken place.
(3) Collapsing and scouring of the medieval fill during pre-boring.

On-Site Works

It was decided to use a specially built kitten rig to pre-bore both the walls and the medieval fill as this machine would install Odex style casing to overcome the anticipated problems. A rotary percussive airflush drilling system was selected, thereby removing any possibility of drill foam or flushing water discolouring the ancient masonry or scouring the embankment fill. The Odex casings, once through the medieval fill, were to be drypacked off at their base and then driven on to achieve an adequate set using an internal steel mandrel hoisted by a purpose built hydraulic winch.

On commencement of the drilling, it was discovered that the rubble fill, which had been pre-grouted, and the embankment fill stood open with very little scouring and hence Odex casing was not required. The piles were therefore drilled and cased immediately and after being driven to the required set were reinforced with a single T16 bar and grouted up into the walls using a high-strength grout placed through an air operated pump and mixer.

Result

The whole operation was completed with the minimum of disruption to the adjacent residents and completed within the contract period allowed.

Case study 8
A traditional underpinning solution

This case study outlines an underpinning scheme that utilised the traditional underpinning technique that has been used for centuries. Unlike many other underpinning schemes that are used to repair failed foundations this particular scheme was proposed to facilitate the construction of a two storey car parking basement beneath the original ground floor of an office refurbishment scheme in the centre of Birmingham.

Design

The bulk of the underpinning was to provide both vertical support to the facade of a seven-storey building and also to provide temporary lateral support while basement construction took place. The remainder of the underpinning was to provide the same vertical and lateral support to a listed banking hall within the body of the building which had to be retained and refurbished. The buildings behind the facade, except the banking hall, were completely demolished prior to underpinning and support was provided to the facade by an external network of structural steelwork which also housed 12 large temporary offices.

On-site works

In all, 200 underpinning bays were constructed under both the facade and the banking hall. The bays were an average of 1.2 m in length and up to 1.2 m wide beneath the corbelled brick foundations, which were at varying levels, corresponding to external pavement levels on two sides. Bay depths varied from 2 m to over 6 m under the facade walk, and between 6 m and 7 m beneath the banking hall walls.

The underpinning bays were constructed in a predetermined sequence to allow sufficient time for curing and for subsequent application of drypack and self-levelling grout between the tops of the underpinning bays and the underside of the corbelled brickwork.

All underpinning bays were constructed in single lift concrete pours, with the inner face of the underpinning vertically aligned with the internal face of the facade walls and with the external face of the banking hall walls. Shuttering was used to full height, strutted back within the working excavation.

Lateral support system

As an alternative to a massive structural steelwork support system designed to be erected as excavation of the basement progressed, a ground anchor system was proposed and subsequently utilised. Coupled with the construction of the deep traditional underpinning bays was the progressive installation of ground anchors at up to three levels (in the deepest underpinning bays).

The permanent works

Basement excavation was carried out once all underpinning bays were completed, as was the ground anchor installation. Once excavated the entire facade and banking hall were supported up to 7 m above formation level which comprised sandstone rock. Reinforced concrete basement construction took place within the confines of the support system.

Result

The underpinning contract was completed in under 16 weeks, shaving well over 60% off the original time period. The work proved satisfactory and the car park was constructed safely.

Case study 9
The use of a jack piled raft

There are numerous areas in Norwich where town fill (i.e. ash bricks, site debris from old demolished buildings) has built up over several hundred years, overlaying the Norwich Craig (sands and gravels) which in turn overlays soft putty chalk substrata. It is these conditions which prevail beneath the three-storey rear addition to a shop unit adjacent to the river Yare in Norwich. A specialist contractor was invited to prepare a scheme for underpinning this building which had undergone severe structural distress. A solution was required which would enable the building to be supported down into the chalk without causing any further disturbance to it or adjacent buildings.

Design

The limitation of avoiding disturbance precluded the use of any type of driven pile and the prevailing soil conditions dictated that a displacement pile was the most suitable. It was proposed that a new reinforced concrete raft slab be constructed to replace the existing floor and would be needled into the existing external and internal walls, with polystyrene formers cast into the perimeter edge beam to enable 150 mm square precast piles to be jacked in through the raft after the concrete had reached its required strength.

On-site works

Prior to construction of the raft the building was tied in with steel straps at first and second floor levels. Great care had to be taken during the construction of the raft to ensure minimal further deterioration of the building.

On completion of the RC raft, it was left to cure until the concrete test cubes indicated the concrete had attained sufficient strength to enable the piles to be installed. The piling operation was carried out in a sequence designed to avoid destabilising the building. This was achieved by jacking in the piles alternately from opposite sides of the building. On achieving the required capacity the pile heads were left approximately 75 mm above the underside of the raft, the connection of the pile to the raft being made with in situ concrete.

Result

The work stabilised the building, enabling restoration work to be carried out safely and without further movement.

Case study 10
Piled foundations for an industrial site

This contract, carried out in 1988, was for the installation of 210 No driven tubular steel piles to carry a working load of 350 kN with a design factor of safety of 2.0. The piles were required to be installed with headroom restricted to 5.2 m. The original programme was to install the piles over a period of 14 weeks, working weekends only.

Design

There was limited site investigation in the areas of the factory where the piles were to be installed so Bullivant installed probe piles in advance of the main piling to establish pile diameter and lengths needed to carry the required loads. These probes were monitored by engineers from both the Ford Motor Company and consulting engineer Posford Duvivier. Four No probe piles were installed at various locations and tested on installation using a pile driving analyser. From the results of these probes it was decided that the most economical design to suit the ground and limited headroom conditions was a heavy duty driven steel tubular pile of 220 mm diameter driven to set between 13 m and 14 m.

On-site works

With the co-operation of the Ford Company and the use of the three fork lift mounted rigs, the piling was completed in 8 days by working two 10 hour shifts during the Christmas plant shut-down. The average pile depth was 13.2 m.

Result

The work enabled the main contractor to complete the installation of a new production line more quickly than originally planned.

Case study 11
Piled foundations on an open site

It was proposed by the Stora Group of Sweden in June 1991 to construct a rail transit terminal in order to ship a variety of freight through to the centre of London, particularly in connection with delivering large paper rolls to the publications industries in Wapping. This called for 1402 piles to carry safe working loads of 200–770 kN at depths of 8–9 m. Installation had to be accomplished in 9 weeks.

On-site works

On-site works consisted of the installation of driven precast concrete piles (250 mm and 300 mm diameter), taking safe working loads of up to 375 kN, founding at a depth of approximately 8.5 m in the underlying Woolwich and Reading beds. Driven steel cased tubular piles (240 mm diameter) were also used on the project to cater for horizontal loading on perimeter piles, each carrying approximately 50 kN. Because of an existing live British Railways track running adjacent to one side of the structure, limited headroom plant was used to minimise the risk of any piles falling onto the track.

A strict testing regime was required by the client on this project, consisting of both static and dynamic testing methods. Both methods gave good correlation, with settlement at proof loads (1.5 × SWL) being in the order of 7–8 mm.

All piles were installed to a predetermined calculated set utilising the long established Hiley Formula, with hammer energies being confirmed by the pile driving analyser. The project was completed ahead of schedule, using two Bullivant tracked rigs with 5.2 T hammers. Out of approximately 1400 No piles, only three were broken due to obstructions within the ground.

After the main project had been completed several other smaller projects were carried out on this site, with over 12 000 m of piling finally being installed on this large contract.

Result

The work was completed satisfactorily on time and the average pile depth was 8.5 m.

Appendix B
Comparative costs

The factors shown in this table are based on the cost of a 1 m deep traditional mass concrete underpinning solution (= 1.0).

Underpinning system	Settlement scheme	Heave scheme
Traditional, 1 m deep (baseline)	1.0	1.2
Traditional, 2 m deep	4.5	5.6
Traditional, 3 m deep	8.0	10.0
Traditional, 4 m deep	15.0	17.0
Pad and beam, 3 m deep	4.0	4.6
Pad and beam, 4 m deep	8.0	9.0
Pad and beam, 5 m deep	12.0	13.0
(100 mm diameter piles, 6 m deep)		
Augered pier	1.3	2.0
Piled raft	1.8	1.9
Angled pile	1.8	—
Pile and beam	2.0	—
Twin knuckle	2.0	—
Driven pier	2.0	—
Twin angle piles	2.0	—
Cantilever pile and beam	2.0	3.0
Jack-down pile	0.33	—

Note: The current actual rates for 1 m deep traditional and 100 mm × 6 m driven cased piles should be ascertained at any given time before applying the factors shown. These approximate comparisons are for guidance only.

Pile only rates compared

This comparison of piling costs is based on the price per metre for 100 mm diameter driven cased piles ($= 1.0$).

100 mm driven cased = 1.0 (baseline)
150 mm driven cased = 1.5
200 mm driven cased = 2.0
150 mm augered   = 1.2
250 mm augered   = 1.3

Further reading

Further information of relevance to those involved in underpinning can be found in the following sources.

Building Research Establishment Digests

BRE Digests covering related topics include:

No 240 *Low rise buildings on shrinkable clay soils: Part 1.* Building Research Establishment, Watford.
No 241 *Low rise buildings on shrinkable clay soils: Part 2.* Building Research Establishment, Watford.
No 242 *Low rise buildings on shrinkable clay soils: Part 3.* Building Research Establishment, Watford.
No 251 *Assessment of damage in low rise buildings with particular reference to progressive foundation movement.* Building Research Establishment, Watford.
No 274 *Fill – Part 1: Classification and load carrying characteristics.* Building Research Establishment, Watford.
No 275 *Fill – Part 2: Site investigation, ground improvement and foundation design.* Building Research Establishment, Watford.
No 276 *Hard core.* Building Research Establishment, Watford.
No 298 *The influence of trees on house foundations in clay soils.* Building Research Establishment, Watford.
No 313 *Mini piling for low rise buildings.* Building Research Establishment, Watford.
No 318 *Site investigation for low rise building: Desk studies.* Building Research Establishment, Watford.
No 322 *Site investigation for low rise building: Procurement.* Building Research Establishment, Watford.
No 344 *Simple measuring and monitoring of movement in low rise buildings – Part 1: Cracks.* Building Research Establishment, Watford.

No 348 *Site investigation for low rise building: The walk over survey.* Building Research Establishment, Watford.

No 353 *Damage to structures from ground borne vibration.* Building Research Establishment, Watford.

British and American Standards

Among relevant Standards are:

ASTM D1586–84 (re-approved 1992) *Standard Test Method for Penetration Test and Split Barrel Sampling of Soils.* American Society for Testing and Materials.

BS 1377: 1990 *Methods of Test for Soil for Civil Engineering Purposes.* British Standards Institution, London.

BS 5387 *Trees in Construction.* British Standards Institution, London.

BS 5930: 1981 *Code of Practice for Site Investigations for Low Rise Buildings.* British Standards Institution, London.

BS 8044: 1986 *Code of Practice for Foundations.* British Standards Institution, London.

Other publications

Attewell, P.B. & Taylor, R.K. (1984) *Ground Movement and Effects on Structures.* Surrey University Press, Glasgow and London.

Barnbrook, G. (1981) *House Foundations for the Builder and Building Designer.* Cement and Concrete Association Publication 48048.

British Geotechnical Society (1989–90) *Geotechnical Directory of the UK.* British Geotechnical Society/ICE, London.

Building Regulations (England and Wales) (1994) HMSO, London.

Bullivant Limited, Roger (1992) Technical Data Sheets. Roger Bullivant Limited, Drakelow, Burton on Trent.

Cambefort, H. (1977) Principles and Application of Grouting. *Quarterly Journal of Engineering Geology*, 10, 53–95.

Clayton, C.R.I., Matthews, M.C. & Simons, N.E. (1995) *Site Investigation*, 2nd edn. Blackwell Science, Oxford.

Cutler, D.F. & Richardson, I.B.K. (1989) *Tree Roots and Buildings*, 2nd edn. Longman, Harlow.

Healy, P.R. & Head, J.M. (1984) *Construction Over Abandoned Mine Workings*, CIRIA Special Publication 32. CIRIA, London.

Hunt, R., Dyer, R.H. & Driscoll, R. (1991) *Foundation Movement and Remedial Underpinning in Low Rise Buildings.* Building Research Establishment, Watford.

ICE (1993) *Specification for Ground Investigation*. Institution of Civil Engineers, London.

ICE Ground Engineering Group Board (1987) *Specification for Ground Treatment*. Institution of Civil Engineers, London.

ICE Ground Engineering Group Board (1988) *Specification for Piling*. Institution of Civil Engineers, London.

Institution of Structural Engineers (1988) *Stability of Buildings*. Institution of Structural Engineers, London.

Leornard, G.A. (1962) *Foundation Engineering*. McGraw-Hill, New York.

Moseley, M.P. (ed.) (1992) *Ground Improvement*. Blackie Academic & Professional, London.

NHBC (1991) *Preventing Foundation Failure in New Dwellings*. National House-Building Council, London.

Padfield, C.J. & Sharrock, M.J. (1983) *Settlement of Structures on Clay Soils*, CIRIA Special Publication 27. CIRIA, London.

Scott, C.R. (1980) *Introduction to Soil Mechanics and Foundations*, 3rd edn. E & F N Spon, London.

Thorburn, S. & Littlejohn, G.S. (1993) *Underpinning and Retention*. Blackie Academic and Professional, London.

Tomlinson, M.J. (1993) *Pile Design and Construction*, 4th edn. E & F N Spon, London.

Tomlinson, M.J. (1995) *Foundation Design and Construction Practice*, 6th edn. Longman, Harlow.

Waltham, A.C. (1988) *Ground Subsidence*. Blackie Academic & Professional, London.

Wynne, C.P. (1988) *A Review of Bearing Pile Types*, 2nd edn, CIRIA Piling Guide 1. CIRIA, London.

Index

access 65, 103, 108, 120–21
accreditation 123, 124
adhesion factor 87
adjacent property 111, 112
advantages 52, 53
age, effects of 6, 7
aggressive agents, 57, 77–8, 79, 109
anchors, steel helical 142
angle of draw 21
angle piling 60, 61, 62, 159–60
approvals 109, 110
associations 13
assurance, quality 122–4
auger 90, 91

base bearing resistance 84, 86–7
beam and pier (pad and beam)
 system 54–8, 137–9, 155–6
beam on pile solutions 139–41,
 146–8, 149–50
bearing capacity 83–7, 129–30
behaviour of ground 83–4, 129–30
buckling of piles 71
Building Regulations 104, 109
building types 6–7, 26
buildings, structural behaviour
 of 4–6

cantilevering 108–109, 146–8
case studies 143–67
categories of underpinning 7–8
CDM Regulations 100, 101–103
cementitious grouts 95, 97–8

civil liability 113–15
Clacquage/hydrofracture
 grouting 94
clay soils
 assessment 31
 bearing capacity 51, 84, 86–7
 characteristics 51
 movement potential 16, 17, 129
 plasticity index 129
 precautions 129
 rehydration 131–2
 suitable foundations 127
 and trees 24–7, 129
compaction 133–5
 grouting 94
 mechanical 133, 135
 vibratory 133–5
concerns
 access 65, 103, 108, 120–21
 householders' 117–18, 119
 insurers' 118–19
 noise 104, 120
 quality 121–4
 working space 103, 121
concrete
 durability 57, 77
 flowing 98–9
 ordinary Portland cement 35, 57,
  73, 75, 76, 77, 79, 99
 piles 78–9
 recommendations 75–6
 repairs 35–6
 sacrificial 79

175

Index

continuous flight auger 90
continuous strip foundations 8–9
contracts 112–13
Control of Pollution Act 1974 115
corrosion 78
cracks 27–8, 29–30
  assessment 31–2
  measurement 32–4
  patterns and profiles 29, 30–31
  repairs 32, 34–6
customer care 118–23

damage
  assessment 24–7, 30–34, 37–8
  causes 4, 25, 27–8, 80
  indications 27–8
  measurement 32–4
design 83–7, 109–110
  flowing concrete 99
  foundation 127, 136–42
desk study 38, 41
difficulties associated with underpinning 103–104
distribution of underpinning work geographically 15
drill/auger rigs 90–93
drilled shaft and grade foundations 10–11
driven (displacement) foundations 11–12
drop hammer rigs 89
durability 77, 78, 79
dynamic compaction 133, 135

effects
  of ground movement 4–6
  of moisture variation 3, 16, 18, 129
  of trees 24–7
erosion 23–4
end bearing 83, 84, 86–7
equipment (plant) for underpinning 87–93
  drill/auger rigs 90–93
  drop hammer rigs 89

Grundomats 88
jack-down rigs 93
top drive hammers 89–90

factor of safety 84, 85, 86
  example calculation 87
foundation types 8–11, 127
  continuous strip 8–9
  drilled shaft and grade beam 10–11
  driven (displacement) pile 11–12
  monolithic slab or raft 9–10
foundations, design of 127, 136–142, 165, 166–7
friction resistance 71
frost heave 20
fuel ash 36, 77, 79, 95

ground improvement 94–5, 131–5
ground investigation 42–6, 126
ground movement
  effects of 4–6
  potential 17
  and trees 24–7
grouts and grouting 94–9
  cementitious 97–8
  chemical 98
  Clacquage/hydrofracture grouting 94
  compaction grouting 94
  flowing concrete 98–9
  jet grouting 95, 96–7
  mini post grouted piles 73
  permeation grouting 94
  selection criteria 95–6
Grundomats 88
guarantees 110–12

health hazards 104
heave 15–19, 28, 57, 129
  frost heave 20
helical anchors 141–2
historical development 1–3
householders' concerns 117–18

infilling of piles 77–9

insurance 1, 12–13, 118
insurers' viewpoint 118
innovative foundation
 systems 136–42
isolated spread footings 10

jacks and jacking 22–3, 157–8
jack-down piling 65, 69, 149–50, 163–4
jack-down rigs 93
jet grouting 94, 96–7
Joint Contracts Tribunal (JCT)
 standard forms 112–13

kentledge 109
knuckle 65–6

legal aspects 100–105, 109–15
 civil liability 113–15
 contractor liability 114
 contracts 112–13
 guarantees 110–12
 legislation 104–105
 negligence 113–14
 nuisance 114–15
 Regulations 100–103
load–settlement relationship 83–4

mass concrete (traditional)
 underpinning 50–54, 161–2
 pad and beam (beam and pier)
  system 54–8, 155–6
mini piles, definition of sizes 59
mini piling 70–73, 157–8
 product range 81–2
mining subsidence 20–22
moisture content variation 16
monolithic slab or raft
 foundations 9–10
mortar mixes for repair work 35
movement, rotational 4–6
movement of slopes 23

needle piling 62, 64–8
needle supports, cantilevered 108–109

neighbouring properties 111–12
negligence 113–14
noise 120
nuisance 114–15

ordinary Portland cement 35, 57, 73, 75, 76, 77, 79, 99

pad and beam (beam and pier)
 system 54–8, 137–9, 155–6
Pali Radice (root) system 73–5
permeation grouting 94
permissible load 84, 86–7
pier system knuckle 65, 67, 151–4
pile and beam system 62, 64–8
pile and knuckle system 65, 66
pile and needle solution 144–5
piled foundations 165, 166–7
piling
 cantilever 65, 68
 design 83–7, 109, 110
 knuckle 65, 66, 67
 mini 70–73
 needle beam 62, 64, 65
 pin 62, 63
 pretest 69–70
 root (Pali Radice) 73–5
 traditional 54–8
 usage 79, 80, 81–2
prevention of failures 125–130
precast grade system 137
precast patent flooring 137, 138, 139, 140
pretest piles 69–70
pulverised fly ash (PFA) 36, 77, 79, 95

quality 121–4
 accreditation 123–4
 assurance 122–4
 standards 123–4

raked piling 60–62
rehydration 131–2
remedial work 111–12

repair work to cracks 34–6
  cementitious materials 35–6
  mortar mixes 35
  sealants 35, 36
restoration 119
root piles 73–5
reinforcement 76
  BS 8004 recommendations 86–7

safe load 84, 86–7
safety factors 84, 85, 86–7
services, public utilities 41, 103–104, 116, 121
settlement 6, 14–17
  consolidation 6
  immediate 6
shoring 103, 106–109
  flying shoring 106–107
  needle and dead shoring 105–106
  raking shoring 106, 107
site investigation 38–41, 48, 49
site survey, process 39–40, 125–6
slenderness ratio 84
slope movement 23
soils (*see also* clay soils)
  bearing strengths 51, 84, 86–7
  desiccation 3, 18
  frost heave susceptibility 20
  general characteristics 51
  investigation 38–9
  load–settlement relationship 83–4
  moisture content variation 3, 16, 18
plasticity index 129

precautions 129–30
relative density 46
yield point 83, 84
specifications 75–7
  cement mixes 75–6
  reinforcement 86–7
standard penetration test 45–6
steel reinforcement, BS 8004 recommendations 86–7
strutting 105–109
subsidence 20-23, 130

temporary supports 105–109
top drive hammers 89–90
traditional underpinning 50–58
  mass concrete 50–54, 161–2
  pad and beam (beam and pier) 54–8
tree root barriers 136
trees, effects of 24–7
trench timbering 108

underpinning
  breakdown of solutions 80
  geographical distribution 15
  reasons for 80
  specifications 75–7
  systems 81

vibration 23, 115
vibrocompaction 133–5
vibroflotation 134
void formers 57

yield point 83, 84